MW01640186

It Seems Like Only Yesterday

Bob and Naomi Lenon on their wedding day, August 14, 1951.

It Seems Like Only Yesterday:

Mining and Mapping in Arizona's First Century

◆

Vol 2: Bisbee and Patagonia

Robert Lenon

with Robert and Judith Whitcomb

R. Lenon
2006

iUniverse, Inc.
New York Lincoln Shanghai

It Seems Like Only Yesterday:
Mining and Mapping in Arizona's First Century
Vol 2: Bisbee and Patagonia

iUniverse books may be ordered through booksellers or by contacting:

iUniverse
2021 Pine Lake Road, Suite 100
Lincoln, NE 68512
www.iuniverse.com
1-800-Authors (1-800-288-4677)

Cover photograph by Robert Whitcomb
Cover design by Robert and Judith Whitcomb

ISBN-13: 978-0-595-36149-6 (pbk)
ISBN-13: 978-0-595-67341-4 (cloth)
ISBN-13: 978-0-595-80593-8 (ebk)
ISBN-10: 0-595-36149-8 (pbk)
ISBN-10: 0-595-67341-4 (cloth)
ISBN-10: 0-595-80593-0 (ebk)

Printed in the United States of America

I dedicate this book to my parents, Harry and Edna Lenon, who taught me to know right from wrong, and to my wife, Naomi, who already knew the difference.

The story of engineering, pieced together from dusty manuscripts and crumbling relics, explains as well the state of the world today as all the accounts of kings and philosophers, generals and politicians.

—L. Sprague De Camp, *The Ancient Engineers* (1993)

Contents

List of Illustrations

Foreword

When Patagonia celebrated its Centennial in 1998, we had the privilege of participating in an oral history program organized by the Patagonia Centennial Committee. In the course of that program, we taped interviews with quite a few members of Patagonia's "Greatest Generation" and transcribed the interviews. Two books have already come from these efforts: *Back in Them Days—When Patagonia Was a Mining Town*, by Joker Mendoza, and Volume 1 of Bob Lenon's *It Seems Like Only Yesterday: Mining and Mapping in Arizona's First Century.* This is the second and last volume of the Lenon memoirs. Like Volume 1, it was originally derived from the Centennial interviews. However, it has been expanded to incorporate written essays from Bob's files and assorted stories that he recorded in interview format. Also, Bob shared with us some of the maps from his immense collection. Many of these maps were ones that he had displayed in the Town Hall during the Centennial. These maps portray the Southwest in general, and the Patagonia Mining District in particular, in a different light from the way we see them today. When they were drawn, the struggle between Native Americans and white settlers was ongoing, and explorers' trails criss-crossed the West. The mines of the region had yet to reach their full potential, let alone experience their subsequent demise. It has been a great pleasure to work with Bob, because in the process we have immeasurably increased our understanding of the history of the southwestern United States, particularly as it applies to Arizona. Also, we know a little bit about mining now! If the readers of this book gain a fraction of this broad historical perspective, our enterprise will have been well worthwhile. Sit back, then, and enjoy this second installment of Arizona history!

Robert and Judith Whitcomb, 2005

Acknowledgments

I acknowledge the assistance provided by Mary Whetzel; Ann Caston; Monique Kornell; Diana Hadley; Wilson Graham; Carol Brooks of the Yuma Historical Society; Carrie Gustavson and Jack Riddle of the Bisbee Mining and Historical Museum; and Rachel Stolworthy of the Franklin (New Hampshire) Public Library. The book arose from the Oral History Project organized in 1998 by the Patagonia Centennial Committee, Owen McCaffrey, Chairman. Peggy Flyntz's excellent transcriptions of interviews paved the way to the final form of the book. Finally, without the continuous and loving support of my wife, Naomi, much that I have recounted here could not have happened, let alone be written about.

Introduction

For those of us who become professionals, our lives develop in two stages. The first is growing up and becoming interested in the area of endeavor that we eventually pursue. The second is heading down the path we have chosen, first as a student and then as a gradually maturing practitioner of our trade.

I grew up in Yuma and also spent some time there as a young adult during the Depression. I wrote about those long ago years in the first volume of my life story, published in July 2004. The small town I remember is long since gone, but it was the site of many historic events, and I consider it a privilege to have grown up there. I moved to Yuma when I was 5, arriving in time to watch the placement of the first highway bridge to span the Colorado River. Later, I traveled on the single-lane plank road that crossed the sand hills of southeastern California on the other side of the river. Today, Interstate Highway 8 crosses the dunes along the same route that the plank road followed.

While I lived in Yuma I witnessed other parts of American history. During the World War I years, my friends and I played war games near my home and watched soldiers preparing to leave our little town for the front lines in Europe. I remember General Frederick Funston, who had become famous as coordinator of relief efforts after the San Francisco earthquake of 1906, riding through Yuma on a big white horse. I also remember Gen. Blackjack Pershing and his troops fruitlessly chasing Pancho Villa into Mexico after the latter's famous border raids. By the time I reached Yuma, construction of the first of the mighty Colorado River dams (Laguna) had been completed. When I was 8 years old, the dam notwithstanding, downtown Yuma experienced a devastating flood. The city will flood no more, however, as mighty dams have tamed the waters of both the Colorado and Gila Rivers, which once met at Yuma. The Gila, at one time large enough to form the original boundary between Mexico and the United States, is now essentially nonexistent, and the Colorado is but a trickle. Yuma remains a major transportation gateway to southern California, as it was at the time I lived there, but the burden of cross-country traffic is now shared by several Colorado River highway bridges.

My father held a variety of jobs in his younger days, but he was primarily a railroad man. He worked for many years for the Southern Pacific. So, as a boy I learned a great deal about railroads from him. Also, I traveled more widely than most boys my age because our family had railroad passes. Nevertheless, when I chose my career, I did not follow in my father's footsteps. Instead I became a mining engineer.

I became interested in mining, at least in part, because of the fascinating tales I heard from two of our neighbors, "Arizona Charlie" Meadows and Felix Mayhew. I also knew other old prospectors, some of whom were friends of my father's or associates of mine when I went out hunting gold or other minerals in the desert as a young man. The mountains around Yuma were rich then in enticing minerals and remain so even now. In those days, many—perhaps most—southwestern men dabbled in prospecting or considered doing so. So when I graduated from high school, I enrolled in the mining engineering curriculum at the University of Arizona in Tucson. This volume begins with my days as a student there.

Patagonia was to become my ultimate home, but it took me some time to get there. I had to endure the Great Depression, while learning mining not only at the huge Phelps Dodge mines in Bisbee and Morenci, but also at the bottom of tiny prospect shafts I explored on my own or with friends. There was a war to fight, too, and I served on both the European and Pacific fronts. When the war eventually ended, I came to Patagonia—near which I had done college field work in the late 1920s—to start a business related to mining and surveying. Much to my surprise, almost all the mines in the area had closed. Nonetheless, I was determined to live in Patagonia. As hard as it was to make a living in mining in a town where it barely existed, I managed to do so and, in fact, eventually prospered. In this volume I will tell you how I did it.

Finally, I will share with you my passion for southwestern history. As I accumulated my collection of thousands of maps, I acquired some that were very old—at least by standards of our young emerging nation. The maps showed cross-country foot trails, wagon trails, and the first primitive roads that men traveled with horses. As I studied these old maps, I became more and more interested in men and events portrayed on them. And so I began studying southwestern history, particularly as it applied to southern Arizona.

In the course of these studies, I encountered a story of a remarkable woman. She was described by A. B. Day, of Franklin, New Hampshire, in an article that he wrote for the November 1, 1906, edition of the *Franklin Journal-Transcript.* The woman, Juana Corona, was a Patagonia resident who was considered at that time to be the oldest living person in Arizona. She was said to be 110. Day reported that she was "as sprightly as any ordinary woman of 50 or 60 years," and was "up in the morning at 5 o'clock, making the fire and starting breakfast." Day felt that Patagonia was a "Resort of Long Life." Maybe he was right. At the time of the Centennial in 1998, there were many residents in their eighties who recalled town history as long ago as the 1920s.

I must have drunk enough Patagonia water that the Resort's influence has affected me, as it did Juana Corona. In fact, it has occurred to me that between Ms. Corona and me, we have lived through the vast majority of the years of our young republic. Assuming that her age was correctly given in 1906, she would have been born in 1796. She would have thus missed out on the first 20 years of the Republic. We do not know when she passed away, but I was born in 1908, and became, in a small way, part of Arizona history in 1914, when I moved from Nebraska to Yuma.

So I end this introduction with a toast to Juana Corona and to the long lives of all my readers!

MINING SCHOOL AND MY FIRST JOBS

University of Arizona

By the time I finished high school in 1925, I had long since decided I wanted to be a mining engineer, so to me there was an obvious choice of schools—the University of Arizona (U of A), which had a very highly regarded mining engineering curriculum.[1] It was the toughest program there, but the professors were very good, and I looked forward to the challenge. I knew that if I found the program to be too tough I could always drop back a step and become a civil engineer or other form of engineer. A mining engineer is, after all, just a civil engineer with a carbide lamp in his hand!

As it turned out, I didn't have to switch curriculums. Although I didn't get A's in all subjects, I learned enough to be able to get along handily enough on every job I ever had. I didn't spend all my time studying. We took field trips to mines in Arizona and Sonora, which, after the tough class work, seemed like vacations. And I also enjoyed ROTC horse cavalry training and hiking in the Catalinas.

During my first two years I lived at the Square and Compass House[2] at the northeast corner of Park Avenue and Second Street. Most of the residents were Masons. I was only 16 when I started college, too young to petition for membership in the Masonic Order. But, in addition to accommodating Masons themselves, the house was big enough for DeMolays, sons of Masons, and even a few men without Lodge connections. I was welcomed because I was both the son of a Mason and a DeMolay.[3] The house management was good, and the residents were pleasant and easy to get along with.

---

1. It still does. It is the only PAC-10 school to even have a mining engineering department.
2. The house was formerly the residence of William P. Blake (1826-1910), a pioneering geologist who is credited with discovering the gold region of California that led to the famous 1840s gold rush (Dellachiara, 2004). He discovered the mineral pool at Palm Springs (Anonymous, 2003). One of the first professors at the U of A, he directed the School of Mines there from 1895 to 1905. We were thus surrounded by the aura of a giant.
3. DeMolay is an organization dedicated to preparing young men to lead successful, happy, and productive lives. It is currently for young men aged 12 to 21, although in my time the age requirement was higher. DeMolay develops civic awareness, personal responsibility and leadership skills but combines this serious mission with a fun approach that builds important bonds of friendship among members. There are more than 1,000 chapters worldwide (DeMolay International, 2002).

As the dependent of a Southern Pacific (SP) worker, I had an annual rail pass and was able to get home to Yuma often. As I mentioned in Volume 1, during the summers after my second and third year of college I worked at a service station in Yuma, earning $25 for an 84-hour week. The first summer, before working at the service station, I spent a few weeks in Phoenix visiting Dad, who was running a work train on a line under construction from Buckeye to Wellton. And, as I also mentioned in Volume 1, I got in a little time that summer as an extra in a silent movie filmed in the California sand dunes. I played a lance corporal in the Royal Lancashire Cavalry. The preliminary title of the film was *The Outpost*, but it was eventually released as *The Desired Woman*. The various jobs I have mentioned plus occasional participation in survey parties or map-making projects helped me finish college with some $700 in hand. These days that would be impossible. It is more likely that college graduates will have thousands of dollars in student loans to repay, no matter how hard they have applied themselves.

The Dripping Springs Gold Rush

On St. Patrick's Day in 1927, when I was a sophomore, there was a gold rush near Dripping Springs Wash, northeast of Winkelman. Almost all of my fellow mining students went to be part of the action. I stayed home, however, to honor St. Patrick. You see, St. Patrick was a civil engineer. He is the patron saint of engineers, who always celebrate March 17. In the past there was a misunderstanding. Some say St. Patrick drove *snakes* in Ireland, but they are wrong. What he really drove was *stakes*.

St. Patrick was an engineer, he was, he was.
St. Patrick was an engineer, he was, he was.
For he invented the calculus
and handed it down for us to cuss.
He drove ***stakes*** *in Ireland!*

So I didn't go on the rush, but I heard all about it when the others came back. John Harris, who was a year or two ahead of me, returned at about midnight with some samples. He had studied assaying and was anxious to identify what he had found. He talked the night watchman into letting him get into the assay office and fire up a furnace. In those days night watchmen were a lot less suspicious than they are today! Unfortunately, John didn't know everything about assaying. Well, he kind of did, but in the end he didn't. When you melt down straight gold ore, you can use a process called inquarting. You add a known (weighed) amount of silver to the charge—enough to represent about a quarter of the amount of gold you estimate is in your sample. If you do it right you get a little "nugget" called doré that is about one-fourth gold and three-fourths silver. You weigh that and then you dissolve the silver—"eat it away," was the way we put it—with nitric acid, and weigh what's left—GOLD!

Now Harris didn't have any pure silver, so in with his sample he put a piece of a U.S. dime—no doubt one of the old Mercury dimes, which had been in use for many years at that time.[4] Those dimes were 90 percent silver and 10 percent copper. (Today, of course, there isn't any silver at all in a dime.) But even with 90 percent silver, he'd have weighed all the insolubles, including a fair amount of impurities, as gold. There was a little bit of gold in there—just a smidgen—but he thought he was rich! Some of his pals took a sample downtown to a real assayer, Jacobs of Tucson,[5] a fixture since the 1880s, and he told Harris it ran 25 cents a ton or something like that. So in the end, it was just one more gold rush disappointment.

Field Training at Mowry

Early in 1928, my junior year, the faculty of the Department of Mining Engineering noted that a thoroughly detailed geological study was needed for a large portion of the Patagonia Mining District.[6] So they teamed up with the

---

4. A dime was significant then but not as significant as it was to become during the Great Depression. An article on the Mercury Dime opined: "No song captures the dark spirit of the Great Depression more than 'Brother, Can You Spare a Dime,' written for a Broadway musical. The song, lyrics by Yip Harburg and music by Gorney Harburg, was first recorded by Bing Crosby in 1932, at a time when one out of every four Americans who wanted work could not find a job" (Lewis, 2004).

   Oh, say don't you remember?
   They called me Al.
   It was Al all the time.
   Say, don't you remember?
   I'm your pal.
   Buddy, can you spare a dime?

5. Jacobs Assay Office, which started up in April 1880, still operates at 1435 South 10th Avenue, Tucson.
6. Mining districts in the Southwest were organized by local mine owners, prospectors, and others with interest in their mining potential. The Patagonia District is a richly mineralized area of the Patagonia Mountains, with significant amounts of quartz monzonite and diorite and of granite porphyry. Values of at least nine ore types are present (Mindat.org, 2005).

Arizona Bureau of Mines and set up a six-week summer course in geology and mapping, for which the Bureau was to furnish the money and the U of A the equipment and staff. The Department contacted mining colleges throughout the nation, and six or seven men enrolled. The course was designed to operate for as many as five or six consecutive summers, with the ultimate goal of producing a finished map and a detailed geologic report on the area of Santa Cruz County from Harshaw to Mowry.

When the course opened, there were two full professors, two assistant professors, and a good camp cook, who had learned his trade as a Buffalo Soldier[7] at Fort Huachuca, but I was the only student! I don't know why the others who had expressed interest didn't show up. Maybe they thought it would be too hot in the summer, inasmuch as it's hot in Tucson, and Patagonia is further south. Of course, because I was going to school in Tucson, I knew the Patagonia Mountains, where we would be working, would be COOL, since they were almost a mile high. Actually, people in my hometown, Yuma, think of *Tucson* as cool!

The Mowry U of A field mapping crew, 1929. Author second from right. Student vehicle, with living quarters in background. Photographer unknown.

7. The Buffalo Soldiers were an all-Black unit. In 1866 Congress established two cavalry and four infantry regiments. The majority of recruits had served in the Civil War. In the 1890s they made up 20 percent of all cavalry forces on the frontier, where they were involved in fighting Geronimo, Sitting Bull, and Pancho Villa (International Museum of the Horse, 1995).

We arrived in Patagonia on July 1, 1928, and were greeted by E. F. (Ed) Bohlinger,[8] who was a de facto one-man Chamber of Commerce for the town. He enjoyed this status in part because he was the representative of the estate of the late R. R. Richardson, founder of the town. Also there to meet us was Ray ("Buck") Blabon, an equally prominent Patagonian of the time. These two men made us welcome and referred us to Bert Logan, caretaker of the land and buildings at the then inactive Mowry Mine, south of Patagonia. This mine had become famous in the 1860s, when it was operated by Sylvester Mowry.[9]

Edwin Fay Bohlinger, circa 1920. Photographer unknown. (Courtesy of Arizona State Library, Archives and Public Records.)

---

8. Bohlinger was a sometime State Representative for Santa Cruz County. Bohlinger Park, just west of the Telles Shrine on State Route 82, is named for him. Today the park—more commonly called simply the Roadside Rest area—is a favorite place for birders.
9. Mowry was a West Point graduate (class of 1852) from Rhode Island who had served in the Union forces but was a southern sympathizer. He bought the mine, then called the Corral Viejo or the Patagonia, in 1860 and made a great success of it. In 1862, however, he was charged with supplying lead to the Confederate forces and arrested. He was jailed in Fort Yuma, a post he had once commanded briefly, and his mine was nationalized. He later established his innocence, was released, and went to London, where he died at the age of 38 (Beck, 1998).

After arriving at the mine, we set up camp in a vacant house that had once been occupied by the mine manager.[10] In addition to housing, we were supplied with a new 1928 four-cylinder Chevrolet pickup and a full set of camp equipment, as well as field and office equipment suitable for several survey parties. Within a day, the professors and I were mapping the topography and geology of the Harshaw/Mowry area. We began by establishing triangulation stations on top of eight or 10 of the most prominent peaks of the area. Then we connected them to the nationwide system of the USGS by setting up a theodolite[11] at a USGS triangulation station atop Red Mountain (el. 6,360) just after sunset. We then measured the angle between lamps set over a similar station atop Gold Hill (el. 4,560), north of downtown Nogales, and one of our new stations on a prominent hill at Mowry, close to a mile high.

Although this process was a routine one, the terrain made its implementation challenging. Just before sunset we sent individual men to the two outlying points with Coleman gasoline lanterns. One professor and I left our pickup truck on the mesa above the Patagonia Grammar School and took turns toting our theodolite to the top of Red Mountain. Believe me, that instrument was heavy, especially when we had to carry it through heavy manzanita brush (there was no established trail then). We measured the assigned angle carefully, just at sunset and early twilight. Then we had to take the theodolite back down the mountain in the dark. It took us until midnight to get back to our truck, with only open-flame miners' (carbide) lamps hanging from our belts to light our way.

10. One former resident of the house was William Mitchell, who had been a classmate of mine at the U of A. His dad had been the manager of the Mowry Mine during the World War I years.

11. A theodolite is an optical instrument consisting of a small mounted telescope rotatable in horizontal and vertical planes, used to measure angles in surveying.

Mine superintendent's residence, Mowry Mine, 1929. U of A field vehicle parked in front. Photograph by U of A Professor R. J. Leonard, Ph.D.

On the next day, the Glorious Fourth, we all took time out to attend the annual rodeo and barbecue at the Circle Z Ranch. This event was hosted for many years by the Zinsmeister brothers—owners of the ranch—and the Patagonia Firemen. My shoulders were black and blue for several days from carrying the theodolite, but the festivities helped take my mind off my aches and pains. Through the rest of July and into August we continued with our mapping. Then I went home to Yuma, where I had a job for the balance of the summer waiting for me at a downtown service station.

At the end of my senior year in 1929, I was short three units of electives for graduation with my class, so I took the field training course again to earn those credits. There were four students that year—Dean Crouch, Jack Rulison, Randolph Goodridge, and me. I finished the requirements for my B.S. degree in mining engineering there at Mowry in the summer of 1929, but I did not actually get my diploma until June 1930.

I have been told that our mapping project was eventually finished. In fact, its scope was enlarged by four or five additional summers' work. Nevertheless, as far as I know, the only map that has survived is the topographic map I began to draft the first year. I still have a print that I used for many years in my business in Patagonia.

Working at the Morning Glory Mine

On completion of my studies at the U of A, I landed a job at the Morning Glory, a relatively small mine 4 miles south of Harshaw and 2 miles west of Mowry. At that time, the Morning Glory had almost completed a fine new "selective" flotation mill at the mine camp off upper Harshaw Creek to concentrate medium-grade copper-silver ores from the underground mine. From my summer work in the area, I knew its manager, J. A. Hamilton, who told me that they were going to operate around the clock, seven days a week. I offered to put my freshly honed skills to work for J. A., and he took me on. I moved into the bunkhouse in the mining camp and ate at the in-camp boarding house, run by an excellent cook, Mrs. Joe Kirkendall, whose husband and grown son, Marshall, also worked at the mine; Joe was the repair and handy man. As I recall, the room was free, and board cost $1.00 a day.

At that time, miners—largely Hispanic, as I remember—were being paid what was then a substantial amount—$4.00 or $4.50 per day. Gringo miners (among them two of my U of A classmates, Ben and Roger Peet) drew $5.00. I was a shift boss, and, as such, I started out at $5.50. I spent my first several days helping to put the final touches on the completion of the mill. It had been skillfully designed and well placed on the southeast side of the gulch, with its highest point just below the level of the main adit (tunnel) entrance. This feature allowed the delivery of mine-run ore to the mill's jaw-crusher along the mine's 18-inch-gauge rail system. Ore was brought down the tracks in trains consisting of five one-ton cars, trammed some 2,000 feet (underground and surface) from the ore chutes inside the mine. The cars were pulled by a single horse, accompanied by a teamster on foot. He had an open-flame carbide lamp to light the way during the underground portion of the trip. Upon arrival at the top of the mill, the driver tilted the contents of the end-dump cars, one by one, onto a strong timber grating, called a grizzly. The grizzly had gaps of 8 or 9 inches or so, through which the chunks of ore fell into a large ore bin that took them to the level of the rock crusher. The larger pieces of ore had to be sledged until they were small enough to fall through the gaps. Horse, cars, and driver repeated this sequence, over and over, until "tally" (quitting time).

There were three work shifts, each headed by a shift boss. Miners' orders were "Fill that bin!" Trammers' orders were "Empty that bin!" Ore was hauled only during the 8 a.m. to 4 p.m. day shift, which was headed by Otis Roberson. I was boss of the graveyard shift, from midnight to 8 a.m., and Duane Walsh was the swing shift boss. We each had a helper, as well as a "crusherman," whose job was to break up the larger chunks of ore before we ground it. The mill superintendent, Edwin Clapp, had had a lot of milling experience in this country and in Mexico. He worked alongside the day shift and made any changes and adjustments required for efficiency and recovery of the ore values. He and the shift bosses made entries in a big bound diary at a little stand-up desk for the information of those who followed them.

Leonard Arrandules was my helper on the floor of the mill. I would tell him, "Turn this on." "Turn that off." "Put a little more chemical in." "Watch it." "Be careful." "Start the pump." "Keep an eye on the water." The well had to be pumped every few hours. It didn't produce water very fast, and, as I noted above, we ran the mill continuously. My crusherman was a fellow from Sonoita. He and his big jaw-crusher were up high behind the mill itself. We didn't even see him except at the beginning and end of the shift. In order to get to him from where we worked, we had to go around to the mine entrance, maybe 500 feet, climb up 30 or 40 feet, and then double back to the crusher.

The crusherman took ore from the chute at the bottom of the bin and reduced it to about egg size in a heavy jaw-crusher. Then it flowed into a fine-ore bin by gravity and went on to the big cylindrical ball mill, where it was mixed with water. The mill, turning slowly on its axis, ground the ore mass into particles that were finer than sugar. This grinding was done in part by the larger pieces of ore, but for greater efficiency, numerous steel balls of about shot-put (15-pound) size were added to the charge. The slurry, moving continuously, flowed into large tanks, where small amounts of various chemicals, including potassium cyanide and sodium xanthate, were added. These chemicals, especially potassium cyanide, are hazardous even in very small amounts. But Ed Clapp warned us to be careful, and we were. I never knew anybody who died from cyanide poisoning in a mine. On my shift, I ate a sack lunch several hours after midnight, after checking all the machinery and flow of ore and chemicals. If I had been working with cyanide, I did peel oranges with my teeth—just to be safe! Sometimes we washed our hands in the warm but dirty mill flow. A little bit of dirt never killed anybody! Fortunately, all the chemicals that we used degraded rather rapidly. There is little or no bioaccumulation of cyanide or xanthate that I know of.

Once the ore was in the tanks of the flotation machines, compressed air was introduced near the bottom. This produced a froth that consisted of a myriad of bubbles. The bubbles, with the help of added chemicals and various conditioners, such as pine oil, picked up valuable material selectively and caused it to overflow from the tanks. The valueless gangue remained behind and, after exiting at the low point of the mill, it was discarded. The choice of chemicals used in the selective flotation process determined which components were saved. The amazing part of this process is that the valuable heavy minerals rise, overflow, and are saved, and the valueless minerals stay below and are discarded! The process is complex and depends on the affinity of the scoured surfaces of the fine particles to the froth bubbles. You have to have particles of 0.5 millimeters or less for it to work well. The surface properties of the particles determine which minerals will be picked up in the froth. Froth flotation methods evolved from earlier film and oil flotation methods. The science of mining engineering dates back a long long way (to the 16th century, as I explained in Volume 1), so even in the early part of the 20th century, there was some pretty sophisticated technology on the job.[12]

We all got along very well at the Morning Glory. As a result, we worked so hard that we just about ran the mine's well out of water! Before long, there wasn't enough water to run the mill, and a little concrete dam that had been built earlier in a tributary to Morning Glory Gulch didn't hold much auxiliary water. So all hands, led by old-time millwright John Wall (1861-1959), whom we called "Uncle Johnnie," built a small timber intake in the main gulch and a good-sized flume leading from it, about 500 feet long, three feet wide and 12 to 16 inches deep, to catch and store flood flows from the summer rains. The flume diverted a good amount of flood water to the pond behind the mine's concrete storage dam. Water from this pond enabled us to mill with surface rainwater while the underground water supply built up a bit. In the late summer of 1929, the Company assayer, Ed Eckel (who later rose high up in the ranks of the U.S. Bureau of Mines), quit to go back to the U of A for an advanced degree. For about two weeks after his departure, I was the assayer, running "fire assays" for gold and silver and "wet" (chemical) assays on 20 to 30 samples per day—mainly for copper values. In only two or three months, however, the mill-grade ore gave out. I suppose Ed realized that this would happen. On September 30, 1929, the entire operation was shut down and all employees were laid off.

Wilfley Tables

Even though I worked hard at the Morning Glory, I took every opportunity to learn more about practical applications. One interesting type of small-mine machinery was perfected at Washington Camp, which was only a few miles from the Morning Glory. This was the Wilfley "concentrating table," devised by A. J. Wilfley. In essence, the tables were powered panning machines. Gold in fine-ground "mill heads" (plus any other heavy material, valuable or not), settled first on the table, which was about 14 feet by 5 feet. It had wooden louvers tacked onto linoleum facing. The mixed ores went down a gentle slope as they were nudged along in water. At Washington Camp, little pieces of gold—maybe the

---

12. For details on the froth flotation method see *Handbook of Ore Dressing* (Taggart, 1927).

size of grains of salt—settled out first in the separating process. Galena (lead sulfide) would follow. Pyrite, which is quite heavy, followed. As the ores went down the slope, streaks of purified ore, about an inch wide, formed. All minerals that are heavy have some value, but Washington Camp, as I remember, was basically a "silver-copper camp." Lead and zinc were not particularly welcome.

BISBEE

Introduction

After leaving the Morning Glory job I moved to Bisbee and within two days was on a new job. When I arrived there in 1929, Bisbee, tucked down in a valley and surrounded by extremely steep slopes, was an exciting place. My home town of Yuma had not yet awakened, but my new home was a major mining town and felt more like a small city than a town. I was attracted not only by its relative sophistication but also by its fine climate. During almost a century of mining—from 1877 until 1975—2.8 million ounces of gold, 102 million ounces of silver, and 8 billion pounds of copper were extracted from the surrounding hills (City of Bisbee, 2004). Bisbee's days as an exciting city were numbered, however, as is true of all boom towns that depend on a finite resource. Underground mining in Bisbee ceased in 1974, and pit operations were shut down in 1975.[1]

The hills around Bisbee remain rich in minerals, especially gold and copper, but because the price of copper has been low for some time, it is not feasible to mine there now. The huge Lavender Pit—named for my onetime boss Harry Lavender—is closed, although it is a major tourist attraction.[2] Mining towns once were unattractive and unhealthy, as their smelters spewed toxic fumes into the air, and, for a time, the Bisbee air was not the cleanest in Arizona. But with the smelters gone, Bisbee has become popular as a retirement and tourist area. It is particularly favored by artists and craftspeople.

What drove me from Bisbee was not the town, but the Company! I was never cut out to be a "company man," a trait that dogged me throughout my Army years and that I never shook off until I became my own boss after the war in Patagonia. But, aside from the frustration of working for a large company, I enjoyed Bisbee while I was there and have many fond memories of the seven years I spent there in the 1930s.

---

1. Excellent records of mining operations in Bisbee are available (Bisbee Mining and Historical Museum, 2004).
2. The large Phelps Dodge open pit mine at Morenci, where I also once worked, is still open, however, as are the PD mines at Miami and Tyrone (in New Mexico). A large new PD mine, which can be easily seen from Interstate 19, has since been opened between Green Valley and Sahuarita, south of Tucson.

Working for the Big Mining Companies

Working With Harry Lavender

Only two days after the Morning Glory shut down, I landed another job. U of A Dean of Mines and Engineering G. M. Butler kept track of jobs available in Arizona so as to find work for his graduates. He told me about two jobs—one in Ruby, northwest of Nogales, and one in Bisbee. Bisbee appealed to me more than Ruby, so on October 3, 1929, I hired out to Harry Lavender, who had been a student of Butler's at the Colorado School of Mines years earlier. Harry had worked his way up at Bisbee and by that time was chief engineer for the Calumet and Arizona Mining Company (C&A). He was kind of a particular guy, and he wanted work done right the first time. As a result, he insulted a lot of people, but I got along very well with him, especially when he found out I knew how to run surveying equipment. He called me "Cacahuate," a Spanish word meaning peanut (properly pronounced kakaWAtay, but Harry said kakawotty).

Harry Lavender, circa 1930. Photographer unknown. (Courtesy of Bisbee Mining and Historical Museum, Bisbee, Arizona.)

Occasionally, Harry would get a hurry-up job, and he'd say, "Let's go out surveying." The first time we went out together he took the instrument and I was the helper. After that, I'd take the instrument and he'd act as the helper. You see, when you're surveying, the head man really ought to be "out front" where he can see what's going on. Harry had lots of experience, and if he was ahead he could evaluate the work better than the man back at the instrument. One time he and I were laying out a bypass road on Company property in Lowell, just east of Bisbee, and we weren't watching for traffic as we should have been. Actually, there *wasn't* any traffic except for one car. We were dragging our 100-foot steel tape along on the ground, and a fellow that we both knew drove by and ran over it. We hollered at him, so he jammed on the brakes. He slid to a stop right on top of the tape and broke it in two! If we'd let him go, the tape probably wouldn't have broken, so it really was as much our fault as his.

The "Submerger"

Early in 1931 the two big copper companies in Bisbee—Phelps Dodge's (PD) Copper Queen and C&A—began work on a proposed merger. The Copper Queen had opened in the 1880s and had the "upper end"—the Bisbee end—of a large ore body, where they ran a first-class operation. Early in the 1900s, upstarts from Calumet, Michigan, took over two adjacent claims, calling themselves the Calumet and Arizona. The claims were richer than they could have imagined. They managed their windfall wisely. They were good miners and businessmen, but they took more business risks than Phelps Dodge did. Eventually, they had the Irish Mag and the Cole Mines, and then they opened the Briggs and Junction shafts and connected all these highly productive mines. They were the little guys to begin with, but the ore bodies they had acquired made them big-time operators over the years, and by 1931 they had extensive underground workings.

In the early twenties C&A hit a major ore body—the Campbell—that turned out to run from the 1,500-foot level down past the 2,700-foot level. It was just a big glob, not a vein. And there was a hundred million dollars worth of high-grade ore in that glob! It didn't even have to be sorted or concentrated. They could just load it from the ore chutes into mine cars, hoist it, put it directly into railroad gondolas, and ship it to their smelter in Douglas, 20 miles east.

When I got to Bisbee the Briggs and Irish Mag claims were both virtually worked out. The Junction, however, had a five-compartment shaft and two steam-operated hoists, which made it a major operation. The Junction and the Campbell both bottomed out at 2,200 feet, and they had just been connected by tunneling from both ends. The new Campbell shaft was concreted rather than timbered, as most mines were in those days. Concreting that shaft and buying a large top-of-the-line Allis Chalmers hoist[3] had cost C&A a lot of money. They had invested wisely in equipment and infrastructure, but as soon as they had done so, along came the Depression. So when the hard times hit, C&A had plenty of ore, but, because of their expenditures on mine improvements, they were a little low on cash. Phelps Dodge had money, but it didn't have all that much ore left in reserve. So the two companies stewed up a merger. I always called it the "submerger," because the Copper Queen (with its Morenci and Tyrone operations) had 52 percent of the total assets of the new company, and C&A had 48 percent, counting other places it had operations, such as Ajo and the "85" Mine in New Mexico. Thus, it was quite close to a "fifty-halfty" deal, but Phelps Dodge had a little edge—the crucial edge!

The big stock market bust that signaled the beginning of the Great Depression occurred just a few weeks after my arrival in Bisbee, only a few days before my 21st birthday. Surprisingly, during the first two years of the Depression, although economic havoc was occurring throughout most of the country, things held together at Bisbee as the two companies made preparations for the merger. Once the merger had been finalized in late 1931, however, the "new" management—which was, in fact, the old PD management—laid off almost all the C&A employees—even the company doctors and nurses. The management did keep most of the mine foremen and bosses, because these workers knew the mines. I knew the mines quite well by then, but I was young and unmarried. Big mining companies don't have much heart, but I think they did try to protect families if they could, so I finally got laid off. I was the last of the single engineers to go.

I Went To Work in Bisbee Twice

As I described in Volume 1, after a period of casting about in Yuma during the worst years of the Depression, I became the first man to be hired to work on

---

3. The Allis Chalmers hoist I mentioned above was spectacular and is worth describing. C&A had used a lesser electric hoist and perhaps also a steam hoist to sink their new shaft. The new double-drum hoist had a massive flywheel. It was a solid piece of metal that had been cast to specification at the AC factory. It was about 12 or 14 feet in diameter and about a foot thick. When it got up to speed, there was enough power stored in it—rotating at about 300 or 400 rpm—for it to "coast" long enough to hoist an entire shift of 200-odd men (at 20 to 25 per trip) up out of the mine, even if the power went off. That wheel was so heavy that they had to take one side out of the hoist house and build a standard-gauge railroad spur to it to bring the new flywheel in on a railroad flat car.

construction of the All-American Canal. In March 1935, however, Harry Lavender asked me to come back to Bisbee. By that time, I was sixth from the top on the AAC payroll, but the AAC project was really expanding then, and the higher up I got, the more politics and paperwork I had to do. That kind of thing wasn't my long suit. Never was, never will be! So I was glad to return to Bisbee, even though it meant a little cut in pay. By the time I returned, the operation was entirely Phelps Dodge. There was no trace of the C&A roots of the corporation. I was the only laid-off engineer who returned; the rest had all found work elsewhere. My duties were much the same as before, but my responsibilities grew. One of the important things I was doing was underground surveying and mapping and making connections between tunnels so they'd meet "square on." Or I'd run raises[4]—maybe about 100 feet—up from one level to the next. I worked alone or with a "pickup" miner helper; it was my responsibility to hit a specific spot alongside the workings above. I never missed one. I almost did once, but my helper—an older fellow who worked with me only an hour or two—saw a flaw in my instructions to him, so we were able to correct it in time.

Several times in 1935 and 1936, I was assigned temporarily to the Morenci and Tyrone branches to assist in lawsuit preparation and tailing-erosion problems. People who have visited Morenci in the past decade or so and seen its gigantic open pit may have difficulty believing that mining there in my time was all underground. It remained that way until Phelps Dodge shut down its entire operation early in the Depression. When it started up again in 1937, it became an open pit (surface) operation.

---

4. There were two types of raises. Manways were passages with ladders for workers to go from one level to another. Ore passes were passageways that skirted the main ore body.

The Lavender Pit Mine in about 1953. Photographer unknown. (Courtesy of Bisbee Mining and Historical Museum, Bisbee, Arizona.)

In about 1937 I was made division engineer at the Campbell Mine in Bisbee. While there I wrote and illustrated two articles on mining methods that were published in the third edition of Robert Peele's *Mining Engineer's Handbook* (1938).[5] Something like 10 of the standard books on mining methods published in the period 1930-50 carried copies of one or two of my drawings from the PD files. And I made a lot of "everyday" maps. In 1968 or so, when my son and I visited Bill Little, a longtime engineer friend who was by then general manager at Bisbee, three of the four large maps on the wall of his office still bore my name or initials. It is very satisfying to a draftsman to see his work gain a little permanence and prominence! So much of what we do is rather transitory.

During the years when I worked at Morenci, the price of copper was down, and there were only 42 people on the payroll. Among them were a doctor and a

5. Peele was a professor of mining engineering at the Columbia School of Mines. He recognized the need for a comprehensive text on all aspects of mining engineering and submitted an outline of a proposed book to leading professionals in 1913. Because World War I intervened, the *Mining Engineer's Handbook* was not published until 1918. Known generally simply as "Peele," it would subsequently serve generations of mining engineers worldwide, with periodic revisions (*Mines Magazine*, Spring 2000). By 1941 there was a second volume (Peele and Church, 1941).

nurse, each of whom worked a 12-hour shift. There was a chief watchman,[6] who had one helper. The fire department consisted of company employees with other full-time assignments; they had man-powered hose-carts to serve the camp's fire hydrants, and alarm boxes to rally the staff in the event of fire.

The Chief Clerk had become the Acting Manager. There were three telephone operators who kept the phones manned 24 hours a day. The mining crew consisted of nine men doing day-shift repairs and general maintenance, plus their foreman, Cornishman Nick Chesterfield. One of the crew's duties was tending the impressive precipitation tanks near the portal[7] of the long tunnel that drained the Yankie workings (named for Joe Yankie, a miner who worked there in the 1880s) into Chase Creek below Metcalf. They had laid off all their mining engineers, so they borrowed me from Bisbee whenever they needed one. One of the lawsuits I helped them with involved a claim-jumping incident.[8] Tom Cocks, a Cornish "local," had jumped several Company claims. I surveyed and mapped these claims and laid out a little topographical model of the surface area of the claims. The model was then built by staff members for presentation in court. Because I had made a commitment to be best man at the wedding of a U of A buddy in Yuma, I didn't stay for the trial. But I didn't have to—the Company won, with the assistance of my model.

On a later occasion I was sent to Morenci when some of their old mill tailings were washing into the irrigated farm districts near Safford. PD was in the wrong that time, so they asked me to develop a mitigation plan. With the aid of a couple of local men whom the manager hired as helpers, I surveyed the area; then I designed five masonry dams. One was more than 200 feet long and about 10 feet high, but the rest were much smaller. These dams restrained and stored stray particles of tailings that had been washing off the downstream sides of the dumps and getting into the general drainage.

The Mexican With the Green Mustache

The Junction shaft in Bisbee had powerful pumps that processed water that collected in the upper levels—thousands of gallons of water a minute. In some parts

6. The chief watchman's daughter had been in my graduating class at the U of A.
7. Entrances to adits and tunnels are called portals; openings of shafts are mouths. A timber framework set around the mouth of a shaft is a collar.
8. Jumping a claim is the commandeering of an existing valid claim by someone who superimposes his own claim on all or part of the previous one.

of the mine—those that had not been worked for a long time—maybe 20 years—there was a lot of copper in the water. It was in soluble form. The copper was solubilized mainly when sulfides degraded, producing sulfuric acid, which then dissolved metals to produce sulfates. It was the sulfates, including especially copper sulfate, that ended up in the mine water. A lot of copper was present in the effluent even if the ore was marginal or even submarginal with respect to copper. "Copper water" is hard on iron and steel. It could ruin expensive machinery or structures in a short time if you were careless. Copper water quickly corrodes iron pipe. The corrosion induced by copper is the result of a quantitative chemical reaction—one atom of copper replaces one of iron. In the reaction the iron is solubilized and is carried away, but the copper is deposited in the form of red copper plating. Back then mine cars and wheelbarrows had iron wheels, not rubber tires. Even those heavy cast iron mine-car wheels would become corroded. Water flowing down the passageways of the mine would nibble away at the rails and track spikes—everything that was iron or steel. You wouldn't know there was a weak spot until an ore car wrecked, and by then it was too late!

But that chemical reaction could be put to use. The basic idea was to put a lot of otherwise useless scrap iron in contact with the water to extract as much of the copper as possible before it could get to the iron and steel equipment. This idea was attractive because the "cement copper" produced, although it was merely sludge, had value, once you solidified it. It was like finding money. In the long Metcalf drainage tunnel northeast of Morenci PD constructed long bins that looked a little like horse troughs, but deeper. They'd put "tin" cans (which are actually steel, of course) and other forms of scrap iron into these bins. They would also use lathe turnings, which, if they weren't oily, were perfect for the purpose. But there aren't enough turnings in all of Arizona to take all the copper out of Morenci. No way!

Bisbee's "Junction Shaft" had similar bins. There was an outdoor "stinker" in which they burned the labels and garbage off of the cans. The cleaned cans were then compressed into bales and sent underground. Once the scrap iron was dumped into the bins, it would turn to copper sludge fairly rapidly. Of course, it didn't really *turn into* copper in the sense of transmutation.[9] In the large wooden troughs operated by the Company, the process was aided by small amounts of compressed air whose bubbles scoured the surfaces clean and dislodged the copper sludge as it formed on the surface of the steel. When you worked in those shafts you knew that the copper water was going to eat the nails out of your shoes and make the heels come off! Of course, there was just a tiny little bit of iron in

9. I explained in Volume 1 why true transmutation of metals is, for all practical purposes, impossible.

your shoes. But it is amazing that a big load of iron in a trough can be pretty much eaten away within a week or 10 days! Copper produced with this process was amorphous, not shiny and metallic—it was essentially sludge. It would be deposited at the bottom of the bins, like mud in a stream. When you shoveled the sludge out, you couldn't take a full shovelful—it was too heavy. And to demonstrate how fast the process occurred, the sludge copper-plated the shiny blade of a shovel instantly.

At the Junction shaft there was a PD employee—a Hispanic fellow—who was assigned to clean out the precipitation troughs. It was a dirty job. Someone doing that all day long would get cement copper all over his body. This Mexican[10] had a big mustache that had turned kind of a sick green from the copper. Interestingly, this fellow and one younger lad were the only nongringos that I recall on the staffs of the Junction and the Campbell while I was there. They probably had trouble finding gringos who wanted a green mustache; Hispanics couldn't afford to turn down a job, no matter how much unpleasantness it involved.

As unhealthy as that poor fellow's green mustache may have been, it paled in comparison to the health hazards in the old smelter towns. Nowadays, of course, there are both state and federal standards controlling emissions, but before they were adopted, smelters released sulfur, lead, copper, zinc, and lots of other toxic materials into the atmosphere. Some people in early-day lead smelting towns had yellow skins, I've been told. I'm sure a lot of people died from metal poisoning (the majority of which may have been from lead) they acquired from fumes and ingestion over the years.

There were also other environmental and safety risks in the old days.[11] At the Kofa Mine near Yuma, for example, copious amounts of siliceous dust was generated as they drilled "dry" with solid steel drill bits. The "up-hole" drills, known as "stopers," were called "widowmakers." If you "drilled dry" for a long time, you died at 60. And you worried about working in quartz veins, too. Local small gold miners worked by hand. They would be exposed to hazardous dust, which was generated mostly in the process of moving ore by shoveling or pulling it from timber ore chutes. Of course, back then, most industries had safety problems. Now OSHA has fixed most of it—so much so that, with all the regulations and masks and shields and boots and rubber gloves and the like, it isn't much fun any more. Like smoking, working in dust and fumes was exhilarating—until you died of it!

---

10. In those days, Hispanics, regardless of their residency, were called "Mexicans." The word had no negative connotations at least as far as I was concerned.
11. The safety risks were recognized long ago (Agricola, 1556), as I pointed out in Volume 1.

Bisbee miners in a typical stope, circa 1910. Note the candles and their holders and the 8" by 8" timbering. Also notice that no safety equipment, such as safety glasses, hard hats, or masks, is in evidence. Mining was a very hazardous occupation in those days. Photographer unknown. (Courtesy of Bisbee Mining and Historical Museum, Bisbee, Arizona.)

How's Your Stinker Working?

Reclamation of copper by the precipitation process in Bisbee led to a funny story. There were only one or two clerks in the mine office. All of them were male. There wasn't even a toilet for women in the engineering office. One clerk was assigned to make phone calls at set intervals to the dozens of mine shops or departments and ask such questions as: "How much ore did you hoist?" "How much drill steel did you sharpen?" Once or twice a week, he'd call up and ask the 2,200-foot-level pump crew, "How much water did you pump?" One call was always to the crew that tended the "stinker." The clerk would ask, "How's the stinker working?" "How many hours did you run?" One day, he misdialed and, without paying attention to who answered, he asked his usual question: "How's your stinker working?" Well, he didn't have the stinker foreman on the line; he had the general office. The secretary there was female, and she let out quite a yell!

Hijinks at the Ajo Pit

In the period during which the merger was being negotiated, none of the Bisbee engineers went over to Ajo. Instead, the Ajo representatives brought the appropriate mine records to Bisbee. This was also true of the "85" Mine at Valedón, New Mexico, near Lordsburg. Ajo Manager Mike Curley came to Bisbee and stayed a while to help, and Al Barr, the chief mine engineer at Ajo, also came over several times.

Even though I wasn't at Ajo much myself, I do remember a couple of interesting tales about it. One was about two old-timer friends, at least one of whom had retired from C&A. Long before they retired, they staked lode claims on nonmineralized ground immediately adjoining the major pit at Ajo. The locations the men were holding blocked further southern development of the pit. The old-timers had anticipated this and realized that they had the upper hand. The pit became kidney bean shaped because C&A couldn't legally mine at the middle of its south edge. The two guys finally sold out to the Company (PD by then) sometime during the World War II years. I think they got $20,000 per claim. They could have gotten a lot more than that if they had continued to hold out, but they were both getting on in years and preferred not to wait.

Over the years, the miners told a funny story that made the rounds again and again. Ajo was always an open pit operation. They had standard-gauge rail haulage from the time they "set in" their first steam shovels circa 1914. Steam shovels and steam locomotives were phased out quite a few years ago, but the general method of shifting the track in an open pit has not changed much, despite modernization and mechanization. As mining proceeds in a pit, the face of the "bench" gets too far away from the track for the bucket of the fixed-rail-mounted power shovel to reach it. Then it becomes necessary to shift the tracks sideways 20 to 30 feet in order to put newly blasted ore within reach of the rail-mounted steam shovels.

The old way to accomplish this involved lining up a couple of dozen track hands, each armed with a heavy pointed 4-foot steel bar. The track boss would command "Lay hold!" "Heave!" and each man would take a good bite in the stone ballast alongside the track, with the bar leaning away from the face of the ore. On command, all hands would pry the entire track over an inch or two. They would repeat the procedure over and over, all day long. Once the track was finally about where it needed to be, the crew leveled it, squared the ties to the rails, checked the gauge, and firmed up the grade for full service haulage.

The early Ajo track crews were made up almost 100 percent of Tohono O'odham (in those days called Papago) Indians. They were very good workers in the heat and got along well with everyone. On occasion, however, there was a Mexican crew. Many of the Mexicans had mainline SP section-hand experience. They could hold up their end on track work jobs, too. Their boss knew how to handle them to get the work out: "Hup!" "Heave!" and teamwork would get the job done. A day's work for a day's pay.

One day an all-Mexican crew was on the job, and the boss was called away unexpectedly. A relief man came in. He was a newly hired gringo with good knowledge of track work from somewhere in the northeast, but he neither spoke nor understood Spanish. Even so, the principle was the same, and away they went, bars in hand, first thing in the morning: "Hup!" "Hyo!" After the third or fourth heave, the new foreman was startled to see fully half of his crew lying on the ground doubled up with laughter. They got up and tackled the job again, with even more startling results. This time they were rolling on the ground, practically in hysterics. And so it went, from bad to worse. It finally took "expert" bilingual advice to straighten out what was going on.

There is always a wise guy in the rear rank of a work crew, and a zany Mexican is pretty funny (except to a supervisor trying to get a job done). So here was this rear-rank character wisecracking in Spanish: "*¿Quién es el payaso mas grande en la mina*?" (Who's the biggest clown in the pit?) And the answer, loud and clear, "*Yo*!" (Me). Just time enough to get off another crack: "*¿Quién es el bromista mas condenado en Ajo*?" (Who's the damnedest joker in Ajo?) "*Yo*!" and "*¿Quién es el zoquete mas estúpido en todo Arizona*?" (Who's the dumbest dope in all of Arizona?) with the by then predictable reply: "I am!" It's pretty hard to just keep on working under conditions such as those, and the track shifting got off to an especially slow start. That day was spoken of with a trace of nostalgia at bars around Ajo for many a year thereafter.

Moving Up—and Back Down

At the time I returned to Bisbee in 1935, all of the real old-timers were still on the engineering staff but were starting to retire. In about 1936, I was promoted to Lease Engineer. I took care of the operation of 10 or 12 underground mineral leases in the portions of the camp that had been worked out years earlier as far as Company operations were concerned. Those old workings had been turned over to faithful longtime Phelps Dodge employees for final mop-up operations, scavenging for the last crumbs of values. In a Company mine, of course, any copper that was recovered belonged to the Company. But it wasn't economical for the Company to extract the very last copper or silver values from nearly depleted workings. This created an opportunity for small-time "know-how" operators to make a little money.

In 1937 or '38, copper prices dropped, and leasing operations were curtailed somewhat. It wasn't worthwhile for even single operators to scrounge the last values out of old deposits. My duties were reduced, so I put in a lot of time going through the office's old records and maps, trying to find references to forgotten veinlets or possible spots for the leasers to scratch away at or, better, if I could find the right "spot," to find extensions of old ore bodies thought to have been virtually exhausted.

That didn't look very impressive, I suppose. One day a sub-boss whom we called the "Floorwalker" told the Chief Engineer: "Watch Lenon, he's not really working." Of course, he bolstered his case by deliberately not assigning me anything important to do. I didn't get along with the guy. He was snippy, and I stood up for myself. Finally, it came time to prepare the annual ore-reserve estimate for the Campbell Mine. Neither the man who had been placed in my old job as division engineer, nor the "Floorwalker," nor the new Chief Engineer knew the mine well enough to do it, so they called on me. When I finished the estimate in May 1940 they gave me my notice, claiming they didn't have anything more for me to do.

In my spare time, after finishing my five-and-a-half-day work week, I was working a tiny scheelite mill that I had set up atop the Huachucas, so after they gave me my notice, I just kept on doing that full time. When I was up there I slept in a cabin Rafael Parra and I built on the property of my friend Harvey James, from whom I was leasing scheelite ore—which was piled up in a 100-ton heap—to run through my mill.[12] (See pages 35-37 for more about the scheelite mill and Harvey.) When my work ran out with PD, I could have requested an assignment at another branch of the Company, but I was tired of working for

12. Harvey had built up the grade of the ore by hand panning, recovering the readily separable tungsten values. However, good values remained, which could be recovered with the aid of a machine.

someone else. Besides, I was making 40 dollars a day with the scheelite! I paid 10 percent off the top for royalty, but the rest was mine.

Drilling

In the mines drilling was your bread and butter. That's the way you created tunnels, shafts, stopes, and raises, and that's how you prepared for blasting ore. When you were driving a heading (creating a passageway) up, down, or sideways, you drilled a "round" (that is, a cluster) of 8 or 10 to 15 or 20 deep holes, in a pattern. Your first holes, when blasted, form a "cut" near the middle of the "working face." The first cavity, which served as a pilot hole, might be 18 inches or so in diameter and 5 or 6 feet deep. Such "cut-holes" are shot (dynamited) with extra powder[13] and often a little extra depth and are fired first, to provide an open space for the round to "break to." After the pilot cut hole is made, more holes—perhaps 12 to 18—are blasted. They may be a little shorter than the cut-holes. When they are blasted, they "slab" the "country rock" or ore toward the pilot opening. The "back holes" (at the top or back of the heading) are elevated a little—having been drilled up 5 degrees or so. After all the other holes have been blasted, the "lifters" are shot. Typically, there are three or four lifters. Like the back holes, they are drilled with a slight angle away from each other and down 5 degrees or so. The miners say "the lifters look down," and, of course, the back holes have to "look up." If they don't, the heading will get a bit smaller after each round as it moves ahead. After the debris gets blown out by the lifters' blasts and the smoke and dust clear, the mucking starts in two phases. The first is "mucking back." This step, done by hand, provides space for continued drilling ahead. The compressed air that powers the rock drills and mucking machines helps to clear the dust and smoke away. The second phase of mucking is loading the ore and—separately—the broken waste into cars for the trammers or "motormen" to haul away.

In those days, a lot of drilling was done by hand; it was hard work—challenging and dangerous. When someone hammered on "hand steel," the top of the steel would splay out a little, and sooner or later a little piece or two would break off. The hammersman who was doing the drilling didn't have to worry much, because his eyes were well up above the drill. But the partner who was holding and turning the steel was right at the site of impact. If splinters were flying he

13. Actually, dynamite is not a powder but rather a jelly-like dough.

could get one in the eye. I never heard of anybody losing an eye altogether while hand drilling, but it may well have happened somewhere. Today there are regulations requiring miners to wear safety glasses. In my day, although they were available, we didn't use them much because they get dirty so quickly underground that it's very hard to see through them. How are you supposed to get your work done if you can't see what you're doing?

The Double-Hand Champion Drillers

Drilling was such an implicit part of mining that mining towns had drilling competitions. I had two friends who were double-hand champion drillers. One was Ben Harden, one of my best prospecting partners. The other, Cliff Newens, was a miner with lots of experience. He worked underground for the Shattuck-Denn mine in Bisbee, which had from 150 to 200 miners. That was a respectable-sized outfit in those days. At that time, hand-drilling contests were held on the 4th of July and Labor Day in the Bisbee Post Office plaza. Ben was maybe 10 years older than I, so he would have been about 45 or 50 when they were entering these contests, and Cliff would have been about 60. So they weren't young, but they sure knew how to drill. They'd practice on granite boulders along a road outside of town. First prize was often two or three hundred dollars. That was a lot of money then. Ben and Cliff were usually good for at least second- or third-place money. Even in the smallest contests, they would usually get a $50 prize.

In serious contests, the organizers provided granite that had been specially quarried in Gunnison, Colorado. Roughly squared boulders, weighing as much as a ton or two each, were shipped by railroad to Bisbee (or any other western town that was having a contest).[14] The use of standardized rock made it possible to compare the times recorded for winning performances in different places. The contest organizers would build a high platform, so the audience could see the action easily. Then they would set a big hunk of Gunnison granite on the platform.

---

14. Alongside the flagpole at the old depot in Patagonia is a local contest stone about the size of a 100-gallon container. The irregular faces have been flattened some, so it is essentially polyhedral. You can see a few holes in the top of the stone that they made in a contest where they recorded speed and depth. But that stone isn't granite. It is pretty soft compared to Gunnison granite.

A Fourth of July drilling competition in Bisbee in the early 20th century. In this case, the platform is set upon a flatbed railroad car. Photographer unknown. (Courtesy of Bisbee Mining and Historical Museum, Bisbee, Arizona.)

The "steels" that the contestants drilled with were shaped in a forge especially for competition. They were expensive, because drilling rules required that they be in matched sets of four or five lengths. Each steel was essentially a "chisel" bit, shaped much like a mechanic's cold chisel. A driller would start out with the largest bit—the greatest diameter across the "blade"; it was known as the "starter." This bit might be $1^1/_8$ inches wide and 18 to 24 inches long. The second steel's bit was $^1/_{16}$ inch smaller in diameter than the starter but a good deal longer. They were made so that the bits would "follow" each other in order. The third bit was perhaps 24 or 30 inches long, and the fourth may have been about 36 or 40 inches long.

If you were drilling single-handed, you would hold the steel with one hand and smack it with a four-pound single jack (hammer) held in your other hand. Then you would turn it about a quarter of a turn so as to make a round hole. In double-handed drilling, your partner turned the steel. If the turning wasn't done just right, the hole would be triangular instead of round. Such a hole is a sign that the turner didn't know what he was doing.

Over the years I had a lot of Mexican helpers. The Mexicans made fun of a triangular hole, which they called a *pata chivo*—goat's foot. Inexperienced miners made that mistake often. A hand-driven steel in a triangular hole could easily get "fitchered." That is, the bit could get wedged in tight. Sometimes it would be so bad that you couldn't have pulled it out with a steam hoist. You would have just lifted the whole stone. You could beat on it, sideways, and just keep working on it and eventually get it out, although you might have to drill a hole alongside of it and blast the new hole (lightly). It happened so often that there is a jingle: "Oh, the ground was fitchered and the steel was stuck, and the whole damn mess was all screwed up." There was always a way in the mine to rescue stuck steel, but that doesn't help you if you're in a timed competition.

So, as I said, in double-handed drilling, at least in my day, one guy held and turned the steel for his partner, who was hammering, and after a while the two changed places. The whole thing was really exhausting and, like the day-to-day drilling, dangerous for the holder (turner). The partner who was turning the steel signaled—by opening up his index finger—that he was ready to stand up and "strike" while the driller "turned" for several minutes. That called for one more lick; then the guy who was turning grabbed the hammer and went Bam! Bam! Bam! and if you were listening but not watching, you wouldn't know that they had shifted. They would change positions without missing a Bam! Sometimes they had a timekeeper to tell them when to swap roles. It was against the rules to hit an extra lick or two after the steel-turner had turned the steel loose; that was called "fostering" and drew a penalty.

All drilling contests were timed. When time was called, the judges used a $^{3}/_{8}$-inch rod with a clamp equipped with a thumb screw and a yardstick-like gauge to measure the depths of the holes. They would run the rod down until it hit bottom and clamp the thumb screw down on its iron plate, which was about 4 inches in diameter. Because the surface of the rock was always somewhat irregular, the plate was needed to establish an average position of the surface to use in measuring the depth of the hole. Once a depth was established for one team, the next team had its chance. The *Bisbee Daily Review* (circa 1936) printed a whole page of the "rules of play," which is on file at the Arizona Historical Society in Tucson. But times change. Today, if you look up "double-handed champion" on the Internet, you will find something about a girl who won last year's double-handed yo-yo contest!

Double-handed drilling was not limited to contests—it started in the mines, and there was always a place for it. As late as 1910 they would use two-man drilling teams in the mines if they had hard rock. Of course it was hard work trading positions with an eight-pound double jack. Also, double-jacking could be awkward if the opening wasn't very high. But you had to drill to get the values out, no matter how hard it was. And the old-time miners were up to the challenge. In those days, strength and stamina were valued very highly. Today miners use air-powered machine drills.

Working on My Own

Mining Gold

I grew up with miners and mining all around me in Yuma, and I did quite a bit of prospecting for gold-bearing ore or placer gold during the Depression. As hard work as it is, mining became a habit in those days. There was money in the rock, and getting it out was an intrinsic part of western life. So, not long after I went back to work in Bisbee in 1935, I started mining in my spare time in the Tombstone and Charleston areas and in the Huachucas.

In 1938 my old friend Ben Harden and Tombstone pioneer Billy Nichols and I were mining for gold on a leased property in "80 Canyon" on the west side of the Huachucas, near the Mexican border. The mine had been pretty much worked out, but we knew where some pay dirt remained. The ore was in a visible white quartz vein, part of which was out in daylight, while the rest was in a short adit, underground. Although it just looked like everyday quartz, it carried quite a bit of gold. You couldn't see the gold in it. I don't think it would even have made a "show" in a pan. But it was there. We drilled and mined that by hand, and it was hard work; Ben and Billy did most of the drilling.

We blasted and then wheelbarrowed the ore out to the portal. At the portal we loaded the ore into matched pairs of wooden pack-boxes mounted on burros. We put a shovelful at a time in each box, so as to balance the loads. Trained burros will stand still while you load. You can put up to 100 pounds on each side, so that part was easy. We had 10 burros, and, even better, we had their owner, Rafael Parra, an excellent burro man and a reliable gent from Bisbee. We took a ton of ore at a time—10 burros carrying 200 pounds each—downhill about a third or quarter of a mile to a little platform we built where big "custom" dump trucks could get to it.

We sent two carloads of that gold ore to the Phelps Dodge copper smelter in Douglas. A Bisbee contractor-friend, Jim Maffeo, hauled the ore about 20 miles to the SP's ramps at Bisbee Junction and shipped it by rail to the smelter for us. About 30 days later we'd get our money, which was basically the value minus freight and treatment. Jim charged us $3 a ton for trucking the first carload because he had bid that for the job, sight unseen. But the partners and Rafael Parra had improved the quarter mile of private road so well that Jim told me he'd haul any subsequent carloads for only $2 a ton. Of course, as luck would have it, it turned out that there was only one more carload, but we did get the bargain rate for that.

There were some important tricks for marketing ore. A friend took me aside and asked, "How did you ship that carload of ore the 20-odd miles to Douglas the other day?" "As gold ore, at $2.50 a ton," I said. And he replied, "Why did you do that? The way Bisbee ores are handled, the rate for 'copper ore' is $0.25 per ton." So I billed the second 50-odd-ton carload out as "copper ore" and made us a hefty saving.

Another factor was important, too. It was a good idea to have a representative at the smelter. My Douglas friend Henry Bollweg, Sr.,[15] a state-registered assayer, represented us when the shipment was weighed and sampled at the smelter.

Less than carload quantities of ore could be sold to a local buyer. We took an additional 900 pounds or so of higher grade ore out of the tunnel and sold it separately to longtime Douglas assayers and ore buyers Hawley and Hawley[16] for $500 or $600.

We leased the ground at 80 Canyon and paid a royalty to the owners amounting to 10 percent of each shipment. The original claimant had lived in Freehold, New Jersey. He was long gone, but his family had been paying taxes on the patented claim for a long time, and they wanted some return out of it. I think we could have worked out a long-term deal if we had wanted to. The family would have been reasonable, because we got along very well. I visited them when I went to New York for the World's Fair in June 1939. But after we had worked that hot spot, we were afraid that any more efforts would just be "dead work."[17] So we ended that particular enterprise.

The Mine We Didn't Make Money On

It used to be quite common to work with people you knew and trusted without formal paperwork. In 1939 I leased a mine near Charleston, over by Tombstone, from the Woolerys, who were good friends of mine. I had gone to college with the Woolery children, and they and their parents had staked several claims on BLM land. They had begun working the claims but had discontinued their work when metal prices declined. The Woolerys let us set up our camp near their property at Howell Spring, which was named for a family that had been around southeastern Arizona since the 1880s. Several ranchers who were leasing grazing rights from the BLM were watering cattle there.

---

15. His son, Henry, Jr., was also an assayer. He had been an engineering classmate of mine at the U of A in the mid-1920s. He got his degree in mining engineering in 1929.

16. Hawley and Hawley was a fine long-lived firm that I could always depend on.

17. Dead work is profitless work, such as blocking out ores in veins to be mined later or searching for additional values.

Ben Harden and Billy Nichols were my partners in that venture, too. I put up the money for food and supplies, but each of us had a one-third interest in the enterprise, which we called the Salary Grabber Lease. I got Ben a big old Studebaker to transport our tools and other supplies. We wanted to be good neighbors, so we dug a well about 6 feet deep in a little draw to avoid disturbing stock coming to the spring for water. Then we began to develop the prospect. I was working in Bisbee at the time, but I drove over and helped on Saturday and Sunday afternoons. Ben and Billy weren't there full time either. Ben and his wife had taken over a Bisbee hotel, and Billy was getting a little old to do a lot of windlass work. We sank several shafts on the vein, drilling by hand and blasting lightly. We brought the ore up with a windlass and bucket. Later, we had an old-timer named Matt out there full time. I kept him on the property just for room and board.

The Salary Grabber was primarily a lead mine, at least where we were working. If metal prices had been higher, we could have made money shipping our ore to Asarco's lead smelter in El Paso. At the smelter they pay a price based on tonnage and percentage of lead, but there would be what they called a "deduct" for zinc if their assays showed a significant amount of that metal. Of course, if there had been enough zinc it could have been concentrated at a profit, but there wasn't enough zinc there to bother with. There are other refractory metals that interfere with smelting and that therefore also trigger deducts. So I started out on the venture by taking samples and having them assayed. Also, the Woolerys had data from previous assays, and they shared their information with us. Part of the time we were driving a cross-cut tunnel into the hill, "feeling" for a parallel vein. Matt could do the tunnel work alone because the rock was fairly soft. He'd drill and blast and would tram broken ore or waste rock out of the blast area in a wheelbarrow, all by himself. He could drive the tunnel maybe 6, 8, 10 feet in a week. He had one limitation. The ore and the wall rock of his adit had a high talc content, which held the dynamite fumes and smoke for hours. So he'd let the muck pile stand idle for several days after each round of blasting—from Wednesday to Saturday or Saturday to Wednesday.

It took two of us to work a shaft. I'd drive over from Bisbee after work on Wednesdays. It was 26 to 28 miles each way. This was in the summer, so the days were long enough for me to go over there and still have time to muck out the shaft we were sinking in an ore vein. I'd go down the ladder and load the ore in the bucket, and Matt would pull it to the surface, 75 to 100 pounds at a time, with our windlass. It was hard work for him to pull it up. I had time to rest in the shaft while he was hoisting it, so it wasn't that bad for me. Although Matt was probably about 70, he was stronger than I was. He wasn't the brightest man in the state, but he was a great worker. We didn't sink deep shafts. We went down only about 45 feet, so one man could work the windlass. If the shafts had been much deeper, we would have needed a second man on the other handle of the windlass. Of course, in that case, we'd have used a 150-pound bucket.

In this "sampling operation," we would sink a shaft in the vein, blast, and then move on to work on another; we would just keep on blasting. We stored the ore in piles—we called them "ore dumps"—right at the collars of the shafts or at the portal of our tunnel. We used 4" by 6" timber for the collar of each shaft and built a heavy wooden frame around each collar. The windlass was mounted on 2" by 6" uprights. We lifted the windlass off when we were going to blast because blasting would have been hard on the rope. When we got down about 40 feet and were just using a windlass and bucket, progress got pretty slow. The man on top was working pretty hard, but the mucker down below was just standing there part of the time. It was a lot of work, and we weren't even mining! We were "proving up" or prospecting the claim. We had a contract with the Woolerys to work the mine if we could develop a paying vein. The stuff we were mining would have been considered "mill ore" a couple of years before, but by the time we got it out to daylight the prices had come down. We held on for about a year and a half, but the price of metals didn't rise. We had developed more than a thousand tons of mill-grade ore, but, given the low metal prices, it just sat there. The price of both lead and zinc had to be good to make it an economic enterprise. We figured prices would come back up, and when they did, our pile of potential ore would finally have enough value that we could make money by milling it or selling it to a big mining company. Our lease and option-to-buy called for us to pay 10 percent of the "net returns" when we shipped or sold ore or concentrates. But we never made a cent. I couldn't hold on forever, so I turned it back to the Woolerys, and we all parted friends. Those claims finally did get mined during World War II.

Prices of Metals

One of the worst things about mining—whether for small-scale operations like ours or for big companies—is the way metal prices fluctuate. When things are bad, the companies have to shut down all their marginal mines and even some pretty good ones. When mine owners see that prices—and profits—are going down, they try to hang on, hoping for a recovery. They tend to say, "Well, we'll just shut down and leave a crew to keep the water pumped out and maybe one or two mining crews to run the dead work." If they kept on with such tasks, it would help them to be ready when prices finally improved. Companies that weren't very well off often had to just keep mining, mining, mining, just to make the payroll and keep the pumps going in wet mines. If you had a lot of ore but money was scarce, that's what you had to do.

You couldn't predict the price changes. Of course, wars always create a demand for metals that are crucial for defense, such as lead, zinc, copper, iron, and manganese. In wartime, if strategically important mines weren't making money, the government could subsidize them. The mining company would say, "Well, the price of copper is 27 cents or 32 cents a pound or whatever, and our ore is marginal. We can't make any money on that." Then Defense Department representatives would say, "We need the metal for defense. We'll pay you a 7- or 9-cent bonus per pound of metal. You go ahead and do the best you can, and we'll make up the difference; the government will subsidize you so you get a profit for your 'break-even' ore."

Actually, the best prices I remember were during the buildup for World War II, before Pearl Harbor. At that time, they still had plenty of miners. Eventually, many of the younger ones were drafted. But by that time they had built up reserves. Prices weren't quite as good once they had stockpiled the needed metals. During major wars, they also subsidized crops, like wheat. For example, farmers might say, "We can't make any money on wheat because we've lost too much of our crop in the drought we've been having." Maybe they wouldn't have the capital to invest after a year or two of drought. But the government would say, "Well, we have to keep you going. Here's 3 cents a bushel or 9 cents a bushel" or whatever. I suppose if there was a lot of rain some years that would drive the prices down, too.

Harvey James's Scheelite Camp

As I mentioned earlier, in 1938 and 1939 I leased a scheelite property in the Huachucas from Harvey James. In those days I was a busy man. In addition to my scheelite enterprise, I was also working in Bisbee and mining lead and gold. Scheelite is calcium tungstate. In its pure state it forms tetragonal crystals with attractive colors, so mineral collectors are fond of it. The scheelite that Harvey had was actually high-grade ore that lay in a thin horizontal vein of quartz on the floor of a limestone cave. But, over the years, volcanic dust had settled in the cave. It was the most viscous stuff I have ever seen. If the scheelite could have been removed directly, it could have been concentrated easily. Unfortunately, however, it was covered with 5 or 6 feet of puttylike mud. Every shovelful had to be pried off the shovel blade. Harvey had mined 50 or 60 tons of scheelite by hand. Then he crushed it with a small hand-operated jaw-crusher into pieces about the size of corn kernels. In the final step he concentrated it to double its scheelite content, all in a gold pan.

It had to be more than 70 percent tungsten for salability.[18] It is common practice to use a fast method of concentration to get quick money and stockpile the "rejects" temporarily for further processing later. That's where I came in.

The mine was at the head of Brown Canyon, high in the Huachucas. Harvey dumped the waste out of the mouth of his tunnel. Even today, if you drive past lower Carr Canyon, on Route 92 south of Sierra Vista, you can see a narrow reddish streak of discarded mine waste coming all the way down the mountain. As I mentioned earlier, Rafael Parra and I built a little rock cabin up there for me, and I prepared to mill Harvey's tailings. I set up a small gravity process mill on a ridge, about 100 feet outside the Fort Huachuca Reservation line fence. I had a little power crusher and a small pair of crushing "rolls"; these tools cracked the coarse particles of tailings to about a quarter of an inch so as to release the scheelite from the quartz. Then I concentrated the primary tailings, which were probably 15 to 20 percent tungsten. I used a little gasoline-powered jig that I had designed. I had it built by Charley Corbett.[19] The components were burroed uphill to the site and assembled there.

My scheelite jig, circa 1939. Photograph by Robert Lenon.

18. Harvey determined the tungsten percentage by filling a Hills Brothers coffee can with dried concentrates and weighing it on a kitchen scale. If it weighed seven and a half pounds it would have been 75 percent tungsten.
19. Corbett was a good mechanic and a Bisbee Chrysler Plymouth dealer from whom I had bought three cars over the years. He was another gent with whom I could do business without having anything in writing. It was better then to do business with a handshake with someone you could trust. And it still is.

The jig used rain water. The separation was a gravity process, based on the difference in weight of heavy minerals (which tended to be valuable) and lighter ones (which were usually worthless). The mix of tailings and water would be "jigged" (agitated). My jig looked a little like a washing machine. It had a heavy screen at the top where the ore tailing (converted into mill-ready ore by this time) was fed in. I used, reused, and re-reused that water! I collected it off all of the buildings—my little cabin, Harvey's cabin, and Harvey's storage shed. The water was stored in 50-gallon oil drums set under the eaves of the buildings. We got lots of rain in those two years. Some commercial jigs may have as much as 100-gallon capacities. Mine was tiny; it might have held 5 gallons. I set up a series of six or eight settling troughs, and I would let the silt (gangue) settle overnight and then carefully decant the water for reuse. We were above 6,000 feet in elevation there, and we got 40 inches of rain a year, so there was enough water for a small operation if we conserved it. The product—semirefined scheelite—came out in particles up to pea size. I dried it by draining. When it was dry, we sacked it and burro-packed it to the end of the mine road, and Harvey drove it in to Hawley and Hawley in Douglas.

Riding a borrowed burro at Harvey James's scheelite camp in the Huachucas. Photograph by Harvey James.

The Vein That Was Big End Up

There was a vein near Bisbee that had its "big end up." It was shaped like a big golf tee—wide and substantial on top but narrowing to little or nothing beneath the surface. In those days we would say that veins like that "wouldn't stand blasting." You think you have a good vein, but when you blast it, the values disappear. In a mining district where the big mines have worked the best deposits, there are still little pockets of ore all around that can stand mining for a month or a week or maybe even only a few days—but then: Poof! You have to go find another one. When prospectors got together and told stories, they would often explain why they weren't mining at a site where they had been extracting ore just a short time before. For example, they would say, "It was a good prospect but it went to nothing!" or "We mined it out—it wouldn't stand blasting!" They would joke about their disappointments—what else could they do? But often it really wasn't very funny. If you had your hard-earned money tied up in mining equipment and had put long hours of labor into your work, it wasn't funny at all.

Fluorspar

I once applied for a fluorspar[20] lease. As long as I had water up at Harvey's scheelite mine, I could make $40 or so per day, clear. That compared very favorably to working for the Copper Queen, where I earned $11.50 or so per day. Even that was actually extra-good pay then—miners were drawing only $7 or $8 a day. I got out a lot of good scheelite shipments, but, despite using my rainwater over and over, I eventually ran out. So I nosed around for something else to do while waiting for the summer rains. I drove over to Tyrone, New Mexico, where a friend of mine, PD manager Phil Lynch, had lessees operating a Company fluorspar lease that I had asked about earlier. For some reason, Phil wasn't happy with the lessees, even though they were operating big jigs and turning out first-class commercial fluorspar. If the deposit was mined properly and crushed and concentrated in one of those big "bull" jigs, it could net $100 or so daily for the lessee. Phil told me I could take over the lease as soon as he was able to close out the current one, which was about to expire. There were experienced local men I could hire to handle the work for me. Unfortunately, however, when I finally got the telegram offering me the lease, six months later, I had been in the Army a week or two into a three-year enlistment! I was called out of ranks (in San Francisco) for the telegram, which said, "Lease is open, come ahead." But I had 35 months and three weeks to serve first.

---

20. Fluorspar (calcium fluoride) is used to make enamel and fancy glass.

Bisbee Town

Bisbee Was a Real Nice Place

Bisbee was just about as nice a place as I could ask for, and I really enjoyed working there. The downtown Southern Pacific Depot was within 10 feet of 5,280 feet in elevation, and the pass on the highway going toward Tucson was 6,200 feet. At that elevation you have good mountain air—not too high to get a person out of breath, but clean and crisp. There was a little snow sometimes, even in town. Working conditions were good, too. C&A's Junction and Campbell Mines had "matching" interconnected levels almost half a mile deep. It was a little hot on the bottom levels, but generally it was a better-than-average mine to be working in. And the outside air was cool at night in the summertime. So after a full day's work I could go home and get a good night's sleep.

Phelps Dodge company store, circa 1935. This store burned down in 1938 but was rebuilt the following year. Photographer unknown. (Courtesy of Bisbee Mining and Historical Museum, Bisbee, Arizona.)

The Copper Queen had a Company store in Bisbee where employees could buy on credit. It was well stocked and competitive with other stores in town, unlike stores in mining camps where workers were beholden to the Company. Even so, I don't recall ever patronizing the place except for their soda fountain. The Marne Café was located across the street from the Miners and Merchants Bank and Trust, which had a nice front porch—one that was pleasant enough to hang out on to watch the town go by. The café was run by three Greek partners who were veterans of World War I; they were all American Legion members. Their names were Gus Pagonas; Pete something or other (I don't think I ever knew his last name); and Mike, the cook—he had a long Greek surname that he never used. Gus was the head man, and Mike was kind of a silent partner. You'd never even see him; he spent all his time working in the kitchen. Gus was married, and his wife was "good help." The other two were single. I'd eat at the restaurant often. Also, for 40 cents or so they'd put up a nice sack lunch that would include a big piece of pie and half a head of lettuce and two sandwiches. We usually ate our sack lunches in the Engineering Office. If we took them down in the mine, we'd have to hang them up high on a string or wire, away from any timbering, so the rats couldn't get to them.

There were two banks in Bisbee. Lemuel Shattuck founded the Miners and Merchants Bank and Trust in the late 1800s to provide competition for the Bank of Bisbee, which had been established a short time before. Shattuck had done very well financially in his co-ownership of the Shattuck-Denn Mining Company. Their mine was reputed to contain one of the most copper-rich ore deposits in the United States. Shattuck eventually opened a branch of his bank in Yuma. I first heard of him there (see Volume 1).

Lemuel Shattuck, circa 1915. Photographer unknown. (Courtesy of Bisbee Mining and Historical Museum, Bisbee, Arizona.)

The 20-30 Club

Late in 1934 I had been a charter member of the Yuma chapter of the 20-30 Club. The organization was like the Rotary Club but consisted entirely of men in their twenties and thirties.[21] We had programs and dinners one night each week. I hadn't been in the club very long when I transferred from the AAC office in Yuma to field work. Sunday was the only day we had off, so I couldn't get to the midweek meetings. But when I went back to work in Bisbee in 1935 they were just starting a 20-30 Club there, so I became a charter member of that club, too, and I was able to go to its weekly meetings.

The Bisbee 20-30 Club sponsored various fund-raising events. One time we sponsored a group that was touring the west with 10 or 12 burros, playing donkey baseball. We split the club into two teams and played softball. You led your burro to home plate and batted, just like in a regular baseball game. But as soon as you got a hit, you had to mount your burro and try to get to first base. You had to have the burro with you at every base you reached. You can imagine that there weren't many home runs! All the burros had names, and I drew one called Coffee Nerves. I hit a pitch—pretty solidly as I remember—and jumped on Coffee Nerves, but he was trained to buck, and he threw me off. I landed on my side, and the whole side of my face slid along the ground. There are no soft places in Bisbee; there's just rock or slag or something else just as hard. So I got a pretty bad scrape on my face. And I never did get to first base.

There was another burro that was trained to "run" clockwise—to third base instead of first! Of course, the batter who was stuck with that burro didn't get very far. None of the burros were good base runners. One big Twenty-Thirtian found a way around this nonsense. He picked up his burro and dragged it to first base! That worked pretty well for one base, but those burros must have weighed about 300 pounds. So he didn't get any further than that. I still have a ticket to the burro baseball game, but there aren't many games like that left to go to.[22]

---

21. Today the organization is known as Active 20-30 International. It began in 1922, when two similar service clubs for young men—Active International in Aberdeen, Washington, and 20-30 International in Sacramento, California—were formed. They eventually merged in 1960 (Anonymous, 2004a).
22. Donkey baseball was once a popular fund raiser. In later years, such games became a target for animal rights activists (Spencer, 2000). Nonetheless, the sport still is seen now and then in scattered venues (Staff Writer, 2003).

Ticket to a donkey baseball game, circa 1936. Local businessmen bought tickets from 20-30 Club members and then gave them to their customers.

The artist Ted De Grazia, who was my age, was from Morenci. I knew his family there. He was manager of a couple of movie theaters and was just becoming known as an artist at the time I knew him. Later he became famous, with an art studio in Tucson.[23] Like me, Ted was a charter member of the Bisbee 20-30 Club. Once the whole club was invited to play a round of golf in Bisbee. We only played nine holes. That's the only time I've ever played, and De Grazia was my partner. Neither of us did very well. I didn't know a mashie from a niblick and still don't!

Bisbee 20-30 Club members preparing for a 9-hole golf round, circa 1936. Author seated in front, second from left, with unidentified child. Ted DeGrazia standing, third from right. Photographer unknown.

23. De Grazia died in 1982.

Our 20-30 group went to Douglas once. I can't recall whether we were inaugurating a new club there or were just getting together with an existing one. We had lemonade during the meeting, but at supper time there was beer, and we used the same glasses for it that we had used for the lemonade. I remember mistakenly putting sugar into the beer, and it fizzed all over the place. I have a picture of the whole group of Twenty-Thirtians. I can't remember the names of half of them, though I think I would know any of them if I saw them. Of course, they'd be 85 to 95 years old by now—if they got old as fast as I did!

The Warren District Scientific Society

I have always been a wise guy. One evening in the summer of 1935 I was sitting around with one of the old C&A gang at his house in Bisbee. By then he was getting pretty high up in the Company—Phelps Dodge by that time. We decided to start a "scientific society." It was totally tongue in cheek. We called it the Warren District Scientific Society. The Warren District was important because it included all the Bisbee mines, most notably the Copper Queen and Shattuck-Denn (Titley and Hicks, 1996). We decided we would try to put a piece in the local daily paper and announce that there was to be a scientific program featuring local speakers. One of our "speakers" was to be Jimmy Vercellino, an Italian who was a contract diamond driller. He had done a lot of work for the Company because he hired good men, had good equipment, and knew what he was doing. His job was to determine whether to drive headings to find ore, and if so where to drive them. We decided he could address our "scientific society" on the "Art of Diamond Drilling."

At that time I was rooming with a married couple, Phil and Lola Bowman. Phil was very well thought of on and off the job. At that time he was a "sampler" for the Copper Queen. A mine sampler in those days had a little prospect pick with which he chipped small representative chunks off of the walls of newly driven "prospect drifts."[24] He then sacked them up to be assayed. It might not seem like hard work, but it was actually physically demanding because it required a lot of picking and toting. Phil also had a sideline—he caught rattlesnakes to be used in research at a college somewhere in the East. He'd catch the rattlers—

---

24. Drifts in mines are passageways that are oriented in the same direction as the vein that they provide access to. It was necessary to sample the potential ore from the sides of drifts; when this was being done, the drift was termed a "prospect drift."

mostly green rattlesnakes[25]—in the Huachucas. He had made a wooden shipping box (such boxes were called "Snake Pullmans"), and he'd send the snakes by Railway Express to a friend in Phoenix, who would send them on to the eastern college. So we decided we would "assign" Phil to give a talk on "Rattlesnakes, Their Cause, Prevention, and Cure"! We immediately had second thoughts about that wording because we knew that it would give us away. So we changed it to something not so obviously phony. A joke is a joke, but if you want it to fly, you have to try to be convincing. So we called it "Rattlesnakes and Their Habitat" or something like that. Now that we had a talk on snakes, we wanted one on spiders. There was a company engineer, Wes Kantner, who didn't know anything about spiders. So we announced that he would discuss "The Life Span of Black Widows." This was a play on words, but it requires some explanation.

It seemed that there were only about six Republicans in Bisbee. Actually, there might have been a couple dozen but not very many. Wes Kantner was one of them. The "spiders" part of his supposed talk about "Black Widows" was only implied. At that time there was a little Black woman in Bisbee—a widow. She was about 50 or 60 years old. This woman, known about town as "The Widow Petty," raised chickens, cooked them, and sold the meat on the downtown streets. There she would be on a street corner, wearing a long Indian-style dress or skirt and toting a couple of covered buckets of chicken. After she sold her chicken, she'd go up and down the street and fill her buckets with food discarded at the cafes. The widow was said to be a registered Republican, and one day she waylaid Wes when he was standing on a street corner and talked to him at some length about the Republican Party's current plans—who was going to run, why, what the issues were and so forth. We found out about the conversation and razzed him about it. We built their talk up into a big romance! We kidded him about being trapped by the Black

25. Seventeen rattlesnake species or subspecies occur in Arizona. The Arizona Diamondback (*Crotalus atrox*) is by far the most abundant. The Green Mojave (*Crotalus scutelatus*) is next in abundance; some estimate their population at 10 to 15 percent of Diamondbacks. Thus the snakes that Phil Bowman was trapping were almost surely Green Mojaves, which have a greenish color throughout their range and are among the most venomous snakes in North America. It is less likely that he was trapping Banded Rock Rattlesnakes (*Crotalus lepidus klauberi*), which vary widely in color but tend to be green in the Huachuca Mountains. Because individual snakes may be almost impossible to see if they lie quietly among lichens, this subspecies is called the Green Rock Rattlesnake.

Widow Petty. We hoped Wes's wife took it as the joke it was intended to be, but I'm afraid she didn't!

The office of the daily newspaper, *The Bisbee Daily Review*, was right next to the Marne Café. I found out from talking to a fellow I knew well who worked in the printing office that they knocked off in the front office about dark and started the press at certain times and had a regular late night lunch hour. So I typed up this piece giving five or six phony speakers (actually the speakers weren't phony, but the talks were!) and stuck my head in the newspaper's front office at a time I expected it to be empty. Sure enough, there was nobody there. I laid my copy on the counter, right where news and notices were received during daylight hours. The night shift found it and printed it without thinking twice! Later my printer friend told me, "We did wonder about it a little, but then we decided it looked all right, so we printed it." Well, it didn't look so good to the Company big wheels at all. Half of them were sore because the chief geologist hadn't been "invited" to be a speaker. The geologist himself was insulted, too, and told people about it. Several other people were also insulted because they weren't included in the "plans."

I should have known better than to do that article. I had overlooked the fact that PD owned the newspaper, and I almost certainly would have been in deep trouble if I'd been caught. In fact, I might have been fired. They figured the article had come from the engineering office because the engineers had done other things like that in the past. In short order, the chief engineer came and asked each of the engineers, "Did you do this?" "No." "Did you do this?" Bill, my co-conspirator, also said, "No." He wasn't really lying—I was the one who had done the typing. Fortunately, when the story broke I had just left to work for six weeks or so in Morenci, and by the time I got back the uproar was all over. The boss never did ask me!

Masonic Lodges

While I was in Bisbee, I became active in the Masonic Order—a longtime Lenon family tradition. My father said that my grandfather neglected his family for the Lodge, and I expect that's true. Grandfather Lenon was Master of four different lodges in three states—Indiana, Iowa, and Kansas. He was a 32nd degree Mason. There's a lot of memory work involved in obtaining the higher degrees, but he did all of it. If there was a part in the 14th degree, for example, and the office holder didn't show up or lost his voice or something, they'd send for "Old Bob," and he'd fill in. He knew the material in full detail by heart. For me that was too much. I just have three degrees. While I was in Bisbee I petitioned for membership in the Yuma Lodge,[26] and, having been accepted, I memorized "the work" of my first two degrees. I got my third degree in Yuma in August 1930, while I was home on vacation. In 1935 I joined the Bisbee Lodge. Because the Arizona jurisdiction permits plural membership, I was a member of both the Yuma and Bisbee Lodges for quite some time. I now hold dual membership in the Yuma and Nogales Lodges, but I don't get to lodge meetings much any more. In fact, the last time I visited my home lodge in Yuma was in 1946. Yuma seems farther away than it once did, and I have been down there less than half a dozen times since then.

One of the big differences between towns today and towns years ago is that the Masonic Lodges are disappearing or, at least, are not as influential in the community as they once were. This is true of many of the old service organizations, such as the Lions Club, Kiwanis, and Elks Club. A Masonic Lodge can be very helpful to its members. For example, in ancient times—at least as far back as the Middle Ages—in Europe a businessman could send cash for something in a far-away city and could depend on the transaction being honest if he was dealing with a brother Mason. Dad was a member of the Lodge in Norfolk, Nebraska, before he went to Yuma. As I described in Volume 1, he got to know the members of the Yuma Lodge soon after he arrived in Yuma in 1913. Members then alerted him to local odd-job opportunities. The jobs that he got through his Masonic connections were very important for all of us because Dad wanted to have a good financial base before he moved Mom and me to Yuma.

---

26. The web site of the Grand Lodge, Free and Accepted Masons of Arizona, states that "The mission of Freemasonry is to promote a way of life that binds like-minded men in a worldwide brotherhood that transcends all religious, ethnic, cultural, social and educational differences; by teaching the great principles of Brotherly Love, Relief, and Truth; and, by the outward expression of these, through its fellowship, its compassion and its concern, to find ways in which to serve God, family, country, neighbors and self." It also indicates that Freemasonry has "great respect for religion and promotes toleration and equal esteem for the religious opinions and beliefs of others."

THE PEOPLE

The Bisbee Miners

There were a lot of ethnic groups represented in the Bisbee mines, but most miners were Caucasians. By and large, the mines I was familiar with rarely hired Hispanics, and if they did hire them, they made them speak English. Many engineers, especially among my classmates and friends, learned Spanish when they worked in mines in Mexico or on the U.S. side near the border.[27] Also, in the late 1890s, during and after the Spanish-American War, a lot of engineers had been in the Philippines, where Spanish was—and still is—the spoken "mining language." However, in Bisbee, English was the mining language, so the bosses weren't required to speak Spanish. This was especially true of C&A and PD in Bisbee and the Old Dominion in Globe.[28]

There was only one Hispanic miner that I remember working in Bisbee, and he was just a lad. His English wasn't very good. I remember him trying to tell me that a hose nozzle was broken, but he pointed to the end of the hose itself instead of the nozzle and called it the "point." I never liked the idea of working in a place where some hands could not understand your language. It was a practical thing. How do you explain that you're going to blast at a specific time and make sure that everybody's away from that part of the mine at that particular time? How do you give any signals for anything? My Spanish is not bad. I always called it "working Spanish," meaning that it is good enough for casual conversation. But I always tried not to work with any crews I couldn't explain everything to in English in some detail and be certain it was understood. I always had a good safety record, and I wanted to keep it that way. Later, in my own business in Patagonia we always spoke English. I didn't hire anyone who spoke any other language on the job. In surveying everything has to be "just so" and full lingual comprehension is important. Clients were a different matter. I trotted out my best Spanish for those who wanted to use that language.

In Bisbee, there were quite a few Italians in the mines, although I didn't know many of them personally. I did know Jimmy Vercellino (of Warren District Scientific Society fame), who was an excellent diamond driller. There was also the

---

27. There were often better opportunities in Mexico, hence better pay. I was offered a job as mine superintendent in a silver mine near Carbó, Sonora, early in 1946. At that time I wanted to go 100 percent on my own, so I did not accept the position.
28. On the other hand, the Patagonia-area mines hired largely Hispanics, as did many other mines in Arizona.

Rolle family in Bisbee and Morenci. I had gone to school in Tucson with three of the Rolle sons, who had grown up in Morenci. But the cultural contributions of the Cornish and Serbian miners are the ones I remember most vividly from Bisbee and Morenci.

The Cornish

There are tin mines in Cornwall, England. Maybe as early as the late Stone Age[29] people in that area mined and refined placer tin (known as stream tin) there. Placer tin deposits are a mineral, cassiterite.[30] The readily accessible deposits were formed by settlement of the heavy pure mineral in running water. Mineral material washes away in the narrower parts of streams, where the water runs faster, but where there's a ripple or wide place it tends to stay in the streambed and can be gathered up and concentrated by panning and smelting to produce metallic tin. Discovery of the tin in the Cornish streams led to the finding of deep, rich veins in the hills, which in turn led to large-scale deep mining. The mines in Cornwall—known as wheals, with the word preceding the name (e.g., Wheal Andrew)—are very deep, maybe 2,000 feet or more. By the late 1920s, however, the mines were becoming increasingly expensive to work and were pretty well played out. Even before the mines were exhausted, Cornish miners began to head for the United States. The hard life in their homeland, combined with the promise of the New World, inspired their emigration, which lasted roughly from 1830 to the early years of the 20th century. They settled first in Wisconsin, California, Michigan, Colorado, Nevada, and the Pacific Northwest. Arizona was one of the

---

29. Engineers (DeCamp, 1993) in the past 8,000 years have made rapid progress, in comparison with the previous 200,000 years of human history. The use of stone tools preceded agriculture by many thousands of years. In the waning years of the Stone Age, copper appeared in conjunction with irrigation and agriculture in Mesopotamia. The Bronze Age, which occurred at various times among different peoples, followed the discovery in about 3500 BC that copper was made substantially harder by the addition of tin to form bronze. Use of iron followed in about 1500 BC. Technically, we still live in the Iron Age, although future historians may have modifications they wish to make to that classification.
30. Cassiterite (stannic oxide) is an opaque mineral whose luster and multiple crystal face cause a nice sparkle. Most aggregate specimens show crystal twins. The most productive sources of this tin-bearing mineral today are in Bolivia.

last locations to which the miners went. Apache wars were an important deterrent to early settlement.[31]

The knowledge that the Cornish brought with them to the mines was very valuable to the big Arizona companies. They knew how to mine and concentrate the ores and how to smelt them. Although the smelting of tin is relatively easy compared to that of other metals, it still helped to have workers who had had a lot of experience with it. They knew how to shore up "heavy ground" and protect themselves while they worked, setting in the supporting timber and wedging it firmly in place.

Up to the time in the Depression when PD had to shut down the Morenci operation, all mining there had been underground. The last underground mine foreman at Morenci was Nick Chesterfield, whom I mentioned earlier, in connection with the limited staffing at Morenci in the late 1930s. "Bloody Nick," they called him. His son Philip worked for me, surveying, for the Company, in 1935, and I'd often see Nick "over across." That meant seeing him going home across the Company grounds. Because he was Cornish himself, Nick took care of the other Cornish miners. For example, if a Cornishman needed a little help patching up his house, a man from the mine's carpenter gang would come up to the house and do the work.

Another way in which the Cornish miners were favored was in the matter of firewood. In those underground mines the timbers were large and heavy and often had to be replaced. Even those as large as 12 inches by 12 inches would last only a month or so in some of the "draw-points"[32] where they were drawing the ore down into the chutes. They'd creak and groan—"taking weight," as the miners say—and they'd have to go. I don't know exactly how they replaced them—it certainly can't have been easy—but the cracked and splintered rejected timber was hoisted up "to daylight" to be discarded. They didn't want it to be lying around below because it could catch fire there. It was the best timber there was—Douglas fir, from up in Washington and Oregon. When they brought it up, they stacked it alongside the collar of the downtown Yankie Shaft, where the townsfolk could get it for their fireplaces or wood stoves. If a miner wanted some of the discarded timber, he'd just go ask the mine captain—"Cap'n, Cap'n, Cap'n, could ya..." There was a Mr. Williams there who had been a foreman in the mine but

31. The story of Cornish miners in America has been laid out in an excellent book by A. C. Todd (1995).

32. Draw-points branch off of "raises" driven up to the ore, which is drawn out through chutes to be hauled to the mill ("concentrator").

was not Cornish. The miners often told of a time when Mrs. Williams was outside her Company house chopping wood so she could light a fire and cook a meal. Nick had arranged to have the wood dropped off, free, by a Company laborer. If it had been taken to a Cornish miner's home, it would have been cut to stove size, as well. But the Williamses weren't Cornish, so they had to chop the wood themselves. One of the neighbors, a Cornish gal, came walking by when Mrs. Williams was chopping and said: "Miz Williams, if thee'rt one of we, thee wudna hafta do tha [pronounced thay (that)]." "Just who in hell wants to be one of we?" Mrs. Williams shot back.

Mrs. Williams's neighbor's language was that of all the Cornish miners—the "Cousin Jacks," as everyone called them. This nickname was given to the Cornish in the United States. It came from the Cornish clannishness. If the foreman of a mine was short-handed, a miner would tell him, "I'll just send back to the Old Country for my cousin, Jack." The Cornish dialect is quite different from the usual British English to which we are accustomed. There was a jingle I heard: "By Tre-, Pol-, and Pen-, by these [prefixes] you know the Cornishmen." And I do remember Trengove, Trevethick, Trestrail, Tredinnick, Trelawny, Tretheway, Trezise, Polwheal, and Penzance. The miners talked about their tools and the mining operations in the language they brought with them. They had different names for many things. For example, a "dag"[33] is a miner's axe: "'And me tha bleddy dag, ol' son," they would say. Hand me the axe!

The Cornish in Bisbee often told a story about a fellow who was giving a burial speech in the Cornish dialect for a miner who had died, quite possibly in the mines. The eulogist said that the deceased had been an excellent miner. "Brother Billy, a good man, mighty fine single-'and [hand] 'ammersman, bloody fine double-'and 'ammersman, bloody fine mucker ary 'and afore" [he could shovel well with either right hand or left hand forward]. And right then he saw a mourner about to break down and said, "Come up and 'ave a look at he, afore we nail the bastard down."

33. A dag has a straight handle so it can be used both as an axe and as a hammer to drive wooden wedges.

Serbians

There were Serbians in Bisbee too. They called them "Bohunks"[34] (a pejorative term for Bohemians—people from Bohemia in Eastern Europe). I remember running a surveying transit near Bisbee alongside a lot occupied by Serbians. As I remember, I was surveying land that the Company would lease out to its employees. Favored people—that is, those whom the Company liked—could buy cheap land, sometimes with a house on it. The couple that I was living with, the Bowmans, were such people. The Company sold them and me a house for $1,500. There was a deep abandoned prospect shaft there—the Saginaw Shaft—that had turned out to have no value, and this house had been the caretaker's. The lot that the house sat on was larger than an "in-town" lot—maybe 100 feet square—on the edge of a little community that was mostly Serbian. Like the Cornish, the Serbians brought their culture with them and tended to group together. Not far from the house that we bought from the company was a church—the St. Stephen Nemanja Church. It had a sign on the front that I couldn't read, even when I got quite close; I finally realized that it was written in the Cyrillic alphabet. The church is still there. There are only three other Serbian Orthodox churches in Arizona—two in Phoenix and one in Safford (Serbian Orthodox Church, 2004).

Crazy Horse Pete

Pete Ivanovich—I don't know how he got his nickname—was a longtime Copper Queen miner. He was small but was one of the strongest men in the mines for his size and was a real hard worker. He could run two big "stopers" (pneumatic rock drills) at the same time while his partner brought him sharp drill-steel! He came

---

34. As in America as a whole, it was common then to recognize the ethnicity of groups of immigrants. Immigrants tended to bond by their mother tongue, and their clannishness invited their being labeled. Today, of course, it is not acceptable to label people in this way. It does seem a little ironic that as much as immigrants were labeled, the Statue of Liberty and the inscription on it were always held in high regard as icons of America throughout most of my life. I'm not sure that is true today in much of the United States, political correctness notwithstanding. We are fortunate to live in Patagonia in a town that has always been, as long as I can remember, without such problems.

from Montenegro.[35] Pete couldn't read or write, but he could calculate with pebbles. I never saw him do it, but I was told that he would just squat down and pick up 8 or 10 marble-sized rocks and set them in a row, something like an abacus. He could figure his wages; he could multiply "five days times four and a half dollars a day" using pebbles.

One afternoon I was sitting on Main Street watching the traffic go by, and here came Pete walking along with his wife. She was a daughter of the local professional fortune teller—"Madame" something or other, some French name. She wasn't French any more than I was, of course, but she called herself that. Now I knew Pete pretty well, but I had never seen his wife. She was nice looking. Pete said, "I want you to meet my wife, fine sonofabitch—give-it-me nine kids." She kind of bowed to me, in the old manner—the way you were supposed to do. "Give-it-me nine kids." He was real proud of her.

Pete and his wife once had a "trial separation." He told me about it some time afterward. They didn't exactly live apart during the separation—except at night. You see, their house was pretty good sized, but they had those nine kids, three or four of them quite young, and there wasn't much room for more. It was a "shotgun house"—one you could see straight through from front to rear. There was a dining room and bedrooms and a kitchen. Mama and the little kids slept in the front room—that was the living room. So Pete put a cot out in a little shed-like room in back where they kept wood for their kitchen stove, and he was going to stay away. But after a week or 10 days or so he began to have other thoughts, and one night he decided to visit his wife. He started to crawl on his hands and knees, going through a couple of bedrooms and the dining room, watching out for the chairs. He had to be careful because there were kids sleeping on the floor. The shades were all drawn and it was black as anything, so he was just going by memory. And then suddenly bang! He had hit his head on something, and he reached up to see what it was, and it was Mama. He was heading north, and she was heading south! That was the last time they tried a separation.

35. It has been estimated that between 1815 and 1915 as many as 35,000,000 people left Europe to better themselves in North America, Australia, New Zealand, South Africa, and elsewhere—"a phenomenal army of souls on the march to a promised land" (Todd, 1995). The Cornish, Serbian, and Montenegrin miners in Bisbee were part of that horde.

Sulfide Pete

There was another Pete—Sulfide Pete—who was quite a character. I don't think I ever knew his real name, and I don't know how he got his nickname, either. Like Ivanovich, he had emigrated from the Balkans. He was a real nice looking guy, maybe 35 years old. He was a good worker who had come in from some other camp. They had him working all by himself on the 570-foot level of the Junction shaft. Now at that time there was hardly anything going on above the 1,300-foot level. I don't know if it was mined out or whether there never was much there in the first place. But there were two or three little levels up that high, and he was there by himself, presumably "on contract." On contract he would be expected to make his "standard," which was set by the shift boss. If you made less exploration footage, you might hear about it, but you would still get your pay. However, if you got more, you would get a bonus. Sulfide Pete did well on the bonuses. I barely knew him, so I don't know if he worked alone because he was hard to get along with. Maybe it was because he had gotten fed up working with bum partners who didn't care if they got a bonus or not. Anyway, he was doing pretty well in comparison to other miners, because he knew what he was doing and worked hard.

Being single helped, too, because he didn't have a family to support. Pete would take a shower on Company premises and on Company time and then would walk down to the suburb of Lowell, where our engineering office was. I'd often see him going by. He was always dressed up; he didn't look like a miner at all. People said he had a Hispanic girlfriend. He had advertised in the papers for a "day-and-night housekeeper." It turned out later that he had been successful. Nowadays quite a few people are finding mates by advertising in newspapers or magazines. Sulfide Pete was way ahead of his time!

Bert Hoover and the Highway Inspector

In about 1930, when I was a new hand at the C&A's Engineering Office, I went home to Yuma for Christmas. It's quite a distance from Bisbee—about 350 miles. Bert Hoover was a mining engineer for PD in Bisbee. He was a pretty nice guy in general, but he also had a sour side. He was short a couple of fingers on one hand as a result of having set off a dynamite cap he had picked up when he was a kid. Still, he did good work. I was working for C&A, and he was working for PD's Copper Queen, so we weren't in regular contact. But he knew that I was going to Yuma and offered me a ride because he was headed the same way. He was pleased to have someone to talk to on the trip, and I offered to pay for his gas, so we got along pretty well.

Somewhere along the way there was a roadside quarantine station. State officers were stopping all the cars, probably looking for hoof-and-mouth disease (see Volume 1 for more on the problems this disease caused between Arizona and California). When we got there, an officer said, "I need to inspect your trunk." Bert's car was an old Chevrolet coupe. Trunks on those models didn't have keys; instead, there was a T-shaped wrench that opened them. So Bert reached behind the seat and handed the official this little T handle so he could open up the trunk. But the official said, "Hey, *you're* supposed to open it up." So they got into words. Bert swore at him pretty heavily, and the guy objected to being sworn at. Bert told him, "Well, when a guy is dumb and doesn't know what he's doing, he needs to be sworn at." Bert was in the wrong, but he got away with it. He insisted afterward that if the guy didn't do right, he deserved to be sworn at. I thought for a minute he was going to use his "club"—his damaged hand—on his adversary. But he didn't. And in a little bit we were driving down the road again.

Bert wasn't the only one around to have lost fingers or hands. There were lots of accidents back then. In my college class alone there were three young men, all from different Arizona mining camps, who were short a finger or two. One, who was from Bisbee, had found a nice piece of shiny copper, a blasting cap, when he was younger. He always shared everything with his twin brother, so he took it home to cut it in two with a hatchet, and it went off and took off the tip of his thumb. When he'd stand or walk around, he'd keep that hand in his trouser pocket so the deformity didn't show. There were quite a few men walking around who had lost more than the tips of a finger or two—things they couldn't hide.

The Mini Rebellion

In the late 1920s, when I first went to Bisbee, there was a superintendent of the C&A smelter in Douglas who had a son about my age—about 21. Right around then there had been a little war going on in Mexico, and this guy had a chance to be a hero.[36] One day an American pilot flying for the Mexicans went to drop a bomb out of an open-cockpit plane on a position in Mexico. But he missed, and the bomb landed just inside the United States—on the PD smelter's slag dump! The south end of the dump was right on the border. The way the story went, the

36. This was the religiously motivated civil war called the Cristero Rebellion, which had begun in 1926, after President Plutarco Calles enforced constitutional provisions that limited the power of the Catholic Church, including a ban on religious schools. In response, church authorities ended public religious functions for three years and incited the worst civil upheaval the country had seen since the Mexican Revolution of 1910-20. Fighting finally ended in 1929, after the Cristero rebels were unable to overcome the federal army's superior numbers and firepower (Butler, 2004).

son decided to take action! He took his hunting rifle and sat on the flat top of the dump, in an area where the molten slag had cooled and hardened. I was told that when the sporadic rifle fire in Sonora died down, he took a few shots to sort of liven things up. I don't know if he wanted to shoot at somebody or just shoot. Whichever it was, it didn't sound like a very good idea to me. He didn't have anything to hide behind! There's no reason that somebody couldn't have taken a pot shot at him. Nobody fired back, I guess, because he didn't get hurt.[37]

Harry Wooten

Harry Wooten was one of the most interesting characters in Bisbee. He was originally from Nebraska. He had built a home in Tombstone Canyon, which was the main street in Upper Bisbee. The home itself cost maybe $6,000. But then he built a retaining wall. It was 12 or 15 feet high along two sides of the block and around a corner, and it cost almost as much as the house!

Harry Wooten, circa 1910. Photographer unknown. (Courtesy of Bisbee Mining and Historical Museum, Bisbee, Arizona.)

---

37. It has been stated that the only American casualty on U.S. soil in a foreign war in the 20th century was a lone victim of a stray bullet in Douglas during the Mexican Revolution (White, 2003).

Harry was a fine gent and ran a first-class hardware store downtown. I don't know how he kept his business going because he was so kind-hearted that he would often give stuff away when somebody came along with a sob story. One day a fellow came by who had a patent for an appliance that he wanted to mass produce. So Harry looked at it and asked, "What's this?" "It's a double line; it's a clothesline without pins." It had two lines, one of which had metal clips to catch and hold the laundered clothes. Instead of handling loose pins you'd snap your clothes onto a fastener on the parallel line. The inventor had made a thousand of them or so, and Harry bought maybe 300. He used a pair for display in his store, but I don't think he ever sold a single one!

Harry Wooten (second from left) in his hardware store, sometime during the World War I years. Photographer unknown. (Courtesy of Bisbee Mining and Historical Museum, Bisbee, Arizona.)

Some years earlier Harry had had a Model T Ford. Most people know what the old Model Ts looked like, but nowadays not so many know how they worked. They had three pedals—reverse, brake, and low gear. Years earlier Harry had gotten his car in a real jam on a dead-end Bisbee street. I remember seeing him sitting on a chair at a party showing guests how he contrived to turn his 12-foot car around in a 14-foot driveway. Another time, in similar fashion, he demonstrated how he had been forced to turn around fast because of a grass fire at the dead end of a narrow alley. He was moving his feet the right way, spinning the (imaginary) steering wheel in the proper direction, and looking over his shoulder at the fire. He did all that on a kitchen chair and told the story, all at the same time. It was quite a workout, as good an exercise as you'll see—going through all those driving maneuvers on a kitchen chair. He was in the wrong profession. He should have been in a circus!

Even though he was a nice guy and a soft touch, Harry lost favor with some of the townsfolk after an incident during World War I—the Bisbee Deportation.[38] The incident has a prominent place in the history of labor-management conflicts in the early days of unionization. At that time, as I explained in Volume 1, Phelps Dodge deported several hundred men (including my friend Harvey James) because of their alleged involvement with the International Workers of the World. Harry Wooten was one of the armed PD "deputies" who rounded up the "troublemakers" and herded them into railroad cattle cars to be run out of town. Some of the deportees eventually returned to Bisbee, and they and their families refused to trade with Wooten because he'd been a Company man.

She Thought I Was a Sheepherder!

At the time I returned to civilian life in 1946, my old prospecting partner Ben Harden and his wife were running a respectable little second-floor rooming house in downtown Bisbee, looking down over Main Street. At that time they had a lady dude—perhaps she could be called a "dudess"—staying there. Ben introduced her to me, saying, "Bob and I used to herd sheep together." She was a nice older lady, and her reply was, "Oh, how sweet!" How was I supposed to respond to that? It was just a joke; it wasn't kind, to say the least, to refer to anyone as "a sheepherder" in cattle country. Nobody thereabouts had anything to do with sheep or sheepherders or even wool shirts![39]

Conrad Hilton

Ben Harden was one of my best friends over the years in Bisbee. He had grown up near San Antonio, New Mexico, a little town near the Rio Grande, just east of Socorro. Conrad Hilton, of Hilton Hotel fame, was also from San Antonio, and his father had a little frontier hotel in Socorro. That was the start of it all—and just look now! Ben knew Conrad as a young man. You would think he would

38. The Governor at the time of the Deportation wrote an account of it (Campbell, 1939).

39. At one time there were large herds of sheep in southern Arizona, but by the turn of the century they were mostly gone, replaced in the main by cattle. Cattle are less damaging to rangeland, while sheep are more suited to cooler terrain. They are still raised in parts of Arizona that get significant rainfall (Kasulaitis, 2005). In 1997 there were 70,000 head of cattle in Cochise County and virtually no sheep (Scorecard, 1997).

have been impressed, but he said that Hilton was the most worthless man he ever saw! Once Hilton went hunting with a bunch of the locals, including Ben, and they came back reporting that he was just a parasite who didn't know how to do anything. He wouldn't even pick up a loose piece of wood and bring it in for the campfire. Ben always wondered how he could ever even have survived, let alone make billions in the hotel business.

Mrs. Bowman's Friend Shops for Chairs in Mexico

As I noted a few pages back, I lived for a time with Phil and Lola Bowman in Bisbee. Lola worked as a cashier for one of the Company stores. Once she told me a story about a time when one of her friends was planning a big bridge party and needed about a dozen extra chairs. She went to a chair factory outlet at a Mexican border town south of Bisbee, probably Naco or Agua Prieta. She looked at various models and chose one. "How much are those?" "They're $4." "Okay, can you make me 10 chairs like that?" And the owner/salesman said, "Oh yes. I'll make you a set." "How much you gonna charge me?" He said, "Well, $50." "Wait a minute! Four times 10 is 40." "Yes, but it's a lot more trouble to make them all exactly the same!" I heard that story several times and always thought it was funny, but looking back on it, it does seem the guy had a point. He didn't have an assembly line where everything that came off was the same. I suppose it *was* extra work to make them all alike.

Mike Curley and His New Buick Touring Car

At the time C&A and the Copper Queen were preparing to merge, I was working with Mike Curley, who was manager of the C&A holdings in Ajo. He came over to Bisbee with the Ajo mine's records, and we evaluated C&A and PD ore bodies for the merger. Mike and I got to know each other pretty well.

One day Mike told me about an experience he had had several summers earlier. He and his wife were getting ready to head east, and he had contacted his dealer in Ajo about picking up a new Buick touring car at the factory in Detroit. When Mike and his wife got to the factory, they were accorded VIP treatment. The plant manager asked, "Do you want to go and watch your car being manufactured?" "Sure, we'd like to do that." There were assembly lines in those days at the big U.S. automobile factories, and sometimes customers were invited to watch the work. The Buick people told him, "Okay, your car is going to have the super this and the extra that, and so forth. So just follow yonder car frame all the way along the production line, and at the end, there will be your car." So away they went, with a well-informed guide, to the place that their chassis was just entering the final assembly line. As they followed it slowly through the various steps, Mike would call for an explanation whenever something happened that he didn't fully understand. He watched as they added the wheels, the motor, and the power train and lowered the body carefully into place. At the end, they filled it with gas and put a cardboard box containing the jack and the lug wrench, a tire iron, and a tire pump under the back seat cushion.

All being in order, the car was test driven, the keys were turned over to Mike and Mrs. Curley, and off they drove, highly pleased. Several days later, in Kansas, they had a flat tire. So Mike reached for the box with the tire stuff and it was gone! They both had seen it go in, but it wasn't there! Apparently, somebody had helped himself while they were parked at a hotel along the way. The people of Ajo, like those in Yuma, didn't like closed cars and therefore tended to choose touring cars, which in those days couldn't be closed up. Mike's car didn't have a trunk, and the thief knew enough about Buicks to look under the back seat. And there Mike was, stranded. All the careful monitoring, but he hadn't foreseen that contingency!

A Better Mouse Trap and Other Tales of Harvey James

Harvey James, my scheelite partner, is long gone now—he died in about 1960. But the years that I worked with him in the Huachucas were among the happiest days of my life. I was on my own and didn't have to kowtow to bosses, and I was making more money by processing the scheelite than PD had been paying me. Harvey was quite a guy. As I mentioned earlier, he had been caught up in the famous Bisbee Deportation because he had sided with the union organizers. The camp in the Huachucas was his home in addition to being a scheelite camp. It was at the top of the mountains, at about 6,000 feet. He had lived there for years, as had his parents.

Bisbee Deportation, July 12, 1917. Armband-wearing men deputized by Phelps Dodge corralled almost 1,200 protesting workers and "railroaded" them via livestock cars to New Mexico, where they were released in the desert. The event was a major milestone in the American labor movement. Photographer unknown. (Courtesy of Bisbee Mining and Historical Museum, Bisbee, Arizona.)

Harvey told a lot of stories. One day he told me how he solved some pack rat troubles he had once had. He had some big snap-type rat traps, and at night he would place one on the dining table in his cabin. But the pack rats had figured out how to set the trap off and then eat the bait! They'd keep him awake at night, springing the trap! A friend suggested that he set some *mouse* traps as well as a *rat* trap or two. So Harvey placed several small, unbaited mouse traps on the table close to the rat trap. When the pack rat approached in search of the bait, his tail set off a mouse trap, and, while thrashing around, he got himself into the rat trap. In those days you "jury-rigged" solutions to problems. You didn't have time to wait for an inventor to come along to design a better trap.

Another of Harvey's tales involved a time during the Depression when he and a buddy, "on the bum," were traveling by "side-door Pullman" (boxcar) and wound up in the Third and Townsend Street Southern Pacific yards in San Francisco. They hadn't eaten for a day or two and didn't have any money to buy food. At that time the city was having a crackdown on panhandling, and if they had been caught they could have been thrown in jail for 30 days. Railroad yards were not a particularly good place to be because they were checked not only by the city police but also by the railroad "bulls." Nevertheless, they were hungry, so they devised a plan. Harvey stood at the mouth of a dead-end street, acting as a "lookout," and his buddy "played" an area half a block down the street. If the cops came, Harvey could admit to being a hobo, but since he wasn't panhandling, they probably wouldn't have arrested him. There was no law against being a hobo. Well, the first fellow that his buddy tackled was deaf, and Harvey could hear only the deaf guy's half of the conversation: the exclamations. "TWO BITS?" "THE BULL?" "ARRESTED?" "THROWN IN JAIL?" So the partner slunk away as quietly as he could. What he had said was, "Can you spare two bits, partner?" "TWO BITS?" "Shshsh. The Bull will nail me." "THE BULL?" ("Yeah, I'll be arrested.") "ARRESTED?" ("I'll be thrown in jail.") "THROWN IN JAIL?" ("Exactly.") The deaf gent was hollering, but Harvey's pal was talking very softly. If the railroad bull—or any cop—had heard all that hollering, he would have run them both in. So they had to give up on trying to raise breakfast money in that particular part of town.

The "bull story" was one of many to come out of the Great Depression. More stories of those days follow in the next chapter.

SOME TALES OF THE GREAT DEPRESSION

In Volume 1 I recounted some of my activities in the Yuma area during the Depression, after I was laid off at Bisbee. When I went back to work in Bisbee in 1935, the Depression was still very severe. I was fortunate to get a job. One of the big events of the Depression was the end of Prohibition. Here are some more stories about that era.

Mr. Friedman's Dessert

Next door to the Marne Café in Bisbee was a nice dry goods and women's garment store, run by a Mr. and Mrs. Friedman. One afternoon when I was in the café Mr. Friedman was arguing with Gus. It seems that at lunch time he had had a Blue Plate Special lunch for 35 cents. Now for dessert with that special there was always a little scoop of Jell-O or ice cream—half the size of a regular scoop—and coffee or tea. Friedman left without having his coffee and dessert, but he came back in the middle of the afternoon and wanted them then. Gus said, "What's this?" "Well, I didn't ask for my coffee and dessert at lunch time, and I would like to take a break and eat them now." Gus didn't exactly argue with him, but Friedman didn't get his food, either. Logically, probably Gus should have given in, but he didn't. So Friedman got done out of 11 or 12 cents worth of fare that he was entitled to. Because this was during the Depression, it was a big deal. Every penny was important!

Al Barr and the Time Clock

In 1932, during the time I was away from Bisbee after being laid off, I made a back-road car trip into Mexico, entering at the San Luis station, south of Yuma. My friend Harry Kieling was with me. After crossing the border, we drove 100 miles *in second gear* on a totally unimproved road! That took us all day. The next day we went to the Pinacate volcanic mountain range. When we came back into the United States on March 5, we used the Sonoyta/Lukeville station, south of Ajo. The Mexican border guard asked, "Where you been?" "Came over from Yuma and been in the country three days. We were in the Pinacate volcano country out in the desert and didn't see one person the whole time we were in there."[1] "How long you been here in Sonora?" "Three nights."

---

1. Actually, we had picked up two young Mexicans almost 100 miles east of San Luis, Sonora, but we dropped them off when we turned aside to the volcanic area. They had been on foot for three days and hadn't seen another car. They were carrying water in 10 or 12 soft-drink bottles, attached to a string harness.

When I went in at San Luis I had paid duty on a five-gallon can of gasoline, so I showed him the receipt. Now I hadn't noticed that the San Luis guard had made a mistake on the receipt. He had just spent 28 days using "2" for the month and absent mindedly did so again, even though he should have used "3." But the guard at Sonoyta had no way of knowing that. He naturally thought that the date was February 3, not March 2. He said, "Where you been for 30 days?" "No, it was just three." But he really thought—or at least he said he did—that we had been in Mexico a whole month and accused me of lying about the length of my stay. So he held my car. Technically, Kieling and I were free to leave, but I didn't want to abandon my car, so I stayed there. Kieling caught a ride into Ajo, where he ran into a U.S. Border Patrol officer—Bert Smith—who was a longtime Yuma friend of both our families. He told our story to Bert, who then invited him to stay overnight at his Ajo home. A bit later, Bert saw the chief Mexican border guard running errands in Ajo. He said, "You let this guy's car go. I've known these two lads since they were little kids. If you don't let that car go, I'll collect about a dozen Sonora cars for no good reason. Then we'll see how you do, holding that car."

Really, my $100 car wasn't valuable enough for the Mexicans to keep, even without Bert's threat. So when the chief guard got back to Sonora he told his men to let my car go. By then it was close to midnight. I drove to Ajo, parked downtown, and spent the rest of the night in the car. Everything was shut down. Not long after I parked, along came Al Barr, Ajo's former chief engineer, afoot. I think Al had graduated from the U of A eight or 10 years before I did. As I mentioned earlier, he and I had become friends when he and Mike Curley represented the Ajo branch of C&A during the early merger negotiations in Bisbee. Later, Al was superintendent at Ajo. At that time in the Depression, however, the mines in Ajo were cutting way back on personnel, as they were in Bisbee and Morenci, so Al had lost his job as a mining engineer. Anyway, here he came down the street. Over his shoulder he carried a "watchman's clock"[2] timer. At each of several dozen key points on Company property (it was Phelps Dodge by that time) there was a key to activate the timer, hanging on a chain. Two or three times each night

---

2. The Antiques Digest (2005) states that a watchman's clock is a "Clock to indicate the time at which a night watchman makes his rounds. Invented by Whitehurst of Derby in 1750, whose version had a large rotating disc with pegs round its edge. The watchman struck a lever which pushed in a peg at the time of his visit. In the modern version the watchman carries a small time recorder. There are keys chained in the places he must visit. On his rounds he inserts each key and turns it, which gives a time indication and the key number on a paper tape sealed in the recorder."

Al would make his rounds, inserting each site's key into his timer and turning it to print out on the timer's tape the times he had been there. It was odd; at 3 a.m. there we were—two former PD employees of good repute standing on the streets of Ajo. We had a good "how've you been" visit. As I remember, Al was getting 5 or 6 dollars a day for carrying the night watchman's timer instead of the 16 or 18 he'd have been making as the chief mine engineer. But he had a job and survived the Depression!

The Ice Plant at Ajo

Because Ajo was a Company town back in the Depression, PD was responsible for supplying the townspeople with staples, and ice was one of those. Therefore, it was PD that had to keep the town's ice plant going after the big cutbacks. They worked out a plan. They said, "Look here. We've got a fire department, and it's got one of our locomotives (a switch engine) with a full-time around-the-clock stand-by engineer/fireman (hostler),[3] who keeps steam up day and night. Now we've got to keep the steam up so as to be ready to take our fire-fighting railroad water cars anywhere in our holdings." Because there was a man there all the time, they said, "We'll make him the fire chief." "He already is." "Yeah, but let's station him and 'his' steam producer at the ice plant. That boiler on the steam locomotive can be hooked up with a fast-detaching hose at the ice plant. The fire chief/locomotive engineer can detach it when he gets a fire alarm and has blown the locomotive's steam whistle to alert his fire-fighting crew. There's enough steam to take tank cars and firemen anywhere in camp to fight the fire with its (steam-powered) pump, using the water we'll have in each railroad tank car." So Ajo consolidated its steam generation capabilities in one place—the ice plant—and was able to eliminate the operation that had been used for the locomotive. The three guys that had worked at generating steam for the locomotive were out of a job, but the PD Ajo operation stayed afloat.

The CCC Program

The CCC (Civilian Conservation Corps) program began in 1933, shortly after Franklin D. Roosevelt took office. It took young single fellows who were out of

---

3. In the parlance of that day, a "hostler" was a railroader who acted as engineer and fireman of steam locomotives at terminals, shops, and yards.

work and placed them in rural "work camps." Regular Army personnel—commissioned officers—were assigned to supervise them, and there were Army doctors and dentists in the camps. The CCC enlistees drew the same pay as an Army private—something like a dollar or dollar and a quarter a day. Their meals were provided, and they were issued outdated versions of Army uniforms and fatigue clothing and shoes. They were expected to send $25 a month from their pay home to their families. Hundreds of thousands of young men soon signed up for the program.[4]

The CCC did a lot of good work in Arizona. It was especially active in soil conservation. When I was in Bisbee during 1929-31 the side canyons and hillsides were all bare. Just about all of the trees had been cut before 1900 for mine timber and firewood. After the railroad came in they could bring mine timber from elsewhere. Also the Copper Queen had a big smelter, and its fumes killed off any growth that was left. Because there were no trees or other growth to divert water when it rained, lots of water cascaded down the canyons after storms. Little waterfalls fell 6, 8, or 10 feet, splashing and then running in streams another 10 or 12 feet and splash, splash, splash again. It was quite a sight. The CCC built little rubble-masonry catchment dams with spillways to slow the erosion, and it worked. Looking at those hills now, you can see that the vegetation came back and stabilized the slopes a long time ago.

---

4. Before taking office as President, Roosevelt proposed to recruit thousands of unemployed single young men into a peacetime force—almost an army—charged with fighting against destruction and erosion of our natural resources. Shortly after being sworn in, he called Congress into emergency session to authorize his program. Only 37 days passed from his inauguration on March 4, 1933, until the induction of the first CCC enrollee on April 7. It was a mobilization of men, materiel, and transportation on a scale never before known in peacetime. Some of the Corps' accomplishments included erecting 3,470 fire towers, building 97,000 miles of fire roads, spending 4,235,000 man-days fighting fires, planting more than 3 billion trees, and arresting erosion on more than 20 million acres. The CCC finally came to an end in 1942, as a result of the lack of applicants, desertion, and a high number of enrollees who were able to find better paying jobs. Obviously, at this time early in the war, the "job" many left for was the Armed Services (NACCCA, 2004).

Upper Brewery Gulch in Bisbee, circa 1930. Note the lack of paving and the bareness of the hillside beyond the town. Concrete steps were later built to facilitate access to hillside homes. Photographer unknown. (Courtesy of Bisbee Mining and Historical Museum, Bisbee, Arizona.)

I didn't live in Patagonia when the CCC was active, but I know there were several CCC camps in the general area, including one at the mouth of Flux Canyon, southwest of town. When I came to Patagonia after World War II, I remember seeing the results of a lot of the CCC projects. In fact, some are still in evidence today. Up on the mesa east of town and down in School Canyon there are several earth and rock dams and dikes. Those little dams were built by the CCC for downtown flash flood control. The big "earth-fill" dam across the canyon above Cady Hall—you can see it from our back yard—is 150 feet long and maybe 20 feet high. They built it in part with dirt they dug out of the spillway-to-be. It was quite a job. When the CCC built things, they lasted. It was a pretty good program, overall.[5]

Marathon Dances

In many towns throughout the United States, there were marathon dances during the Depression. A lot of people didn't have jobs, so they entered the dance contests, and the couple that held out the longest won some money. Sometimes the dancers would go on for 30 days and nights—24-7, as they say nowadays. In Bisbee such dances were held in Lowell, a pretty tough part of town right alongside the mines. The Dance Hall was an annex of the White House Cafe. I remember hearing about one dance in Lowell that finally got down from 15 or 20 couples to two after 17 days. Suddenly, after all that time, a dancer named Jane—who was said to be really tough—slapped her partner in the face and walked off! She said her partner had insulted her. Nobody ever found out *what* he could have said to insult her! But whatever it was, it ended the contest.

Prohibition

From 1914 to 1933 there was prohibition in Arizona,[6] but everybody drank anyway. Many people made home brew, and a little "on-the-sly" commercial brew

---

5. My old Yuma friend Harry Kieling was an Army Reserve lieutenant (later captain) on the CCC staff in charge of supplying several CCC camps along the Mexican border. He stayed in the Army and eventually retired as a full colonel after World War II.
6. Asbury (1968) summarized the national status of prohibition in 1919, when the 18th Amendment was ratified: "in January of 1919, thirty-three states were already dry, in addition to Alaska, Puerto Rico, and the District of Columbia," and the Anti-Saloon League said officially that if the local option districts in wet states were included, 95 percent of the continental United States, with 70 percent of the population, was dry. However, only 13 states were *bone* dry: Arizona, Arkansas, Colorado, Georgia, Idaho, Kansas, Montana, Nebraska, Oklahoma, Oregon, South Dakota, Utah, and Washington. The rest allowed the manufacture and sale of liquor under various restrictions.

was also around. I remember that they had Old Sunnybrook whiskey in Yuma, and, as I remember, there was some Mexican booze as well. The commercial booze was "bottled in bond" for export and smuggled north across the line. And there were bootleggers. George Proctor, who now lives in Patagonia, grew up near Madera Canyon, right on the Santa Cruz–Pima County line. When George was only about 5 or 6, his dad distilled mescal from agave plants and would send him out on a horse to deliver jugs of the stuff. He would tell George to go across the valley and leave the jugs alongside a certain tree to be picked up. George notes today that he "went very quietly" on these delivery runs. His family's big sheet-copper still is around to this day (Portillo, 2003).

When I moved to Bisbee in October 1929 to work for Harry Lavender, there was another unmarried engineer, Ellsworth Gundy, who had been working there for several years. He and I were the only single guys in the Engineering Office. When I arrived he was rooming with the son of a homeowner who lived up Wood Canyon, in Upper Bisbee, but this roommate was about to get married and leave town. I was living in the rundown Warner Hotel, so Gundy (he didn't like his given name) asked, "Do you want to move in with me up in Wood Canyon?" And I said, "Sure!" We had room and board at the house. I don't remember what we paid for it. We'd come home from work, eat supper, and go downtown to the movies. One night we came home, and there was a blanket over the window above the kitchen sink. And there was the owner's wife, distilling "mash." She obviously had some expertise; we watched her testing the specific gravity of the product with a "float."[7] I said, "What's happening?" Gundy and the homeowner replied, "Look, we've both got cars and monthly payments to make." The homeowner was a miner. He worked hard, but he wasn't rich. Young engineers weren't making a lot of money either. Gundy said, "Where do you think we've been getting those monthly payments? I told these folks you wouldn't tell anyone what was going on." And I said, "I don't intend to tell on you, I'm just commenting." And he said, "Okay." So they were distilling mash to be sold as whiskey. Two local brothers—miners, as I remember—came by and bought a small wooden keg every month. I believe that it held about 5 gallons. Another local also had a keg on order every month. And they probably gave some, maybe a gallon or so, to the sheriff, just to be sociable.

---

7. The specific gravity (a measure of liquid weight relative to water) was an indicator of alcohol content. Good hooch contained a relatively high content (perhaps 20 percent) of ethanol and was therefore lighter than water.

There were speakeasies too. When I was at the U of A in the 1920s I'd go to a beer joint once in a while. There were little houses down in "Adobetown." Whole families would be involved. They probably paid off the sheriff and the chief of police with a few jugs of beer. Some of the home-brewed beer was pretty horrible, but if the guy who was making it knew what he was doing, much of it turned out okay. It was made with yeast, and some of that would stay in the bottle—that probably didn't do your insides any good. It was available for "two bits" a glass but in those days, coming up with 25 cents was the problem. Men worked for a dollar a day in the Depression and were glad to get that much.

Almost all of the printers at the *Yuma Morning Sun*, where I worked during high school, were drinking guys. Some of them drank a partially dealcoholized wine called Virginia Dare[8] that was available legally at the drugstore as a "tonic." It was alleged to be a "builder-upper." Of course, the medicinal compounds were probably close to zero in their effectiveness. The original good feeling, which made you feel that the tonic really worked, came from the alcohol. But the "drys" eventually pushed through legislation that made the Virginia Dare brewers put a laxative in their product. The winos would get the trots if they drank very much. But they drank it anyway. It ran 12 or 14 percent alcohol. That's what they wanted, and they were willing to trot to get it. It was the same then as it is now: drunks had to have their booze. In those days, if they couldn't get "good stuff" some of them drank wood alcohol, which could cause blindness or have other horrible effects on their health. Sometimes they drank the stuff because they were not very smart, and sometimes it was an accident—for example, if a bootlegger mixed wood alcohol in to extend his product. You could buy wood alcohol legally at the drugstore, and they might combine that with orange juice or some other cover-up. I guess that type of thing still happens: some drug dealers adulterate their products today, I understand.

One Too Many for the Crusherman

I knew somebody who died from drinking the bad kind of alcohol. He had been a crusherman at the Morning Glory Mill in 1929. When I got out of the Army I tried to look up my old friends around Patagonia. I asked, "Whatever became of

---

8. Virginia Dare wine, dating to 1835, was produced from the Scuppernong grape. In 1919, when prohibition was instituted, the alcohol content of the wine was reduced. By 1923 the Virginia Dare Extract Company had been founded. Today it is known for producing a good vanilla extract (Virginia Dare Corporation, 1998).

the crusherman?" "Well, he lives up by the windmill. You know where that windmill is up west of Sonoita? It's right across from the track, the horse track, not the county's fairground track, or the railroad track, but the site is just a half mile or so southwest of Sonoita." So I stopped by to see him. He was in his fifties when he had worked with me, and this was 17 years later. But he said, "Oh, yeah, I remember you." I said, "Well, I heard about your well. Friends say you have the best well in the county." "This well is *the best*," he said. "I don't drink any other water. We have running water in the house, but that's for baths and washing dishes." We went out by the windmill. He had a dipper and a covered wooden barrel to collect the pumped water that he would get direct from the windmill. Extra water ran downhill and watered his garden. And he said, "I just make out. I never did drink running water in the house. It's got an iron pipeline." So I joined him in an H_20 "salud" (salute to health) and I went on my way.

A year or two later my friend came home late. As I heard the story, he'd been drinking, and apparently he couldn't get to sleep. Now this was *after* Prohibition, and good stuff was readily available, but he evidently didn't have any. So he went out in the horse shed and rounded up the bottles there. There was Sloan's liniment, which was 18 percent alcohol, and there was something else that was 14 percent. He drank all of the Sloan's and at least some of the other one. They never figured out which he took the most of, but they were both unsuitable, and he died. He never woke up. Here he was taking such good care of himself—not drinking water out of an iron pipe line and things like that—but he died anyway. I heard that story several times. I don't know why he was drinking that bad stuff. He left a wife, grown kids, and grandchildren behind.

Supervising a Gold Mine

My stay in the Bisbee area ended when I had a chance to go back to one of my old stomping grounds—southern California. By that time it was 1940 and the Depression had lifted a little. There was a mine manager there named Fargo Rose. He had an old prospecting friend who had found a good little virgin gold vein in Long Valley (south of Pine Valley), California, by crawling around on his hands and knees under incredibly dense brush—manzanita, in part—that no one else had ever penetrated. No one had ever suspected the mineralization there, and there wasn't so much as a prospect pit for a mile around until early in 1940. When the mine and little mill got going in a very small way Fargo was made mine manager. I had once worked as assayer-engineer with his mill man, Bob Kelso, at a gold mill near Yuma. Kelso was having recovery troubles. They were not recovering the last 25 percent of the gold values in their mill. He told Fargo, "There's a young fellow from Yuma, living in Bisbee, who solved similar troubles for us in a gold mill near Yuma, in 1932." So Fargo drove more than 500 miles to see me in Bisbee on Labor Day weekend, 1940. He asked me to go with him and lend a hand. I was working at my scheelite lease in the Huachucas, but I dropped everything and went along with him. I was able to solve their problems, and they offered me the position of superintendent of the new gold mine and mill. So I went back to Bisbee, cleared up my business there and in the Huachucas, and returned to California a week or two later to take up the job.

We had a crew of 18 men, mining and milling. We were down about 35 feet in the shaft, just getting into the sulfide zone[9] nicely on about a 30-inch-wide, almost vertical quartz vein, when the World War II draft hit. I had a fairly low draft number and might not have been called up for 12 to 18 months. But I did not like that uncertainty hanging over me, and I felt that I could use my skills in one of the armed services. By that time at the mine I had adjusted the gravity milling equipment and tinkered with our selective flotation "cells" until they were saving well over 90 percent of the gold values, so I felt that I could leave without negatively affecting their operation. Around the middle of December 1940 I quit my job and went home to Yuma for the holidays before enlisting for a three-year hitch in the Army.

The Two Fargo Roses

Fargo Rose is an odd name, of course—so odd that it is even more odd that there could have been two of them. Fargo was born in about 1900 and was raised in San Diego. He was bright and energetic and had been around a good deal. He had been a Marine lieutenant during the middle twenties, when Marines rode the mail trains to foil hold-ups, which had become especially bold along about then.[10] I recall seeing enlisted Marines in uniform on the streets in Tucson, circa 1927, on their time off. Tucson was a rail division point, and I presume it was a layover point in this armed escort program.

---

9. Minerals in deep deposits often occur as sulfides. When these sulfides are exposed to oxygen, in air or water or both, they may be converted to oxides. In many deposits, there is a depth beneath which oxygenation has not occurred. This is the "sulfide zone."
10. In the days between the early Pony Express and stagecoaches and the advent of a good highway system and airplanes, trains were the principal means of transporting mail between cities. Many items shipped had great value. Currency itself was also often sent by train. Naturally, enough, trains therefore were highly attractive to robbers. There was a wave of train robberies throughout the nation in the 1920s (Campbell, 2001).

Fargo told me that, when he was a young fellow in San Diego, he met someone from Fargo, North Dakota, who told him that he knew another Fargo Rose, a girl his age born in Fargo and still living there. So our hero wrote to the female Fargo. They got to be pen pals and seemed to be quite compatible. Eventually, our Fargo went to North Dakota to visit the girl. He got in town fairly early one morning and went out to her house. Daddy Rose answered the door. He was not the friendly type at all but did say that his daughter was away from home but would be back that evening. So our hero strolled back downtown and whiled away the intervening hours in the pool halls. Unfortunately, he also imbibed a few beers to gain the courage to face the ogre again. Unfortunately, when he did return, the father wouldn't even let him in the door. He gave Fargo the old heave-ho, with very little in the way of formalities. Fargo Rose "1" never got to meet Fargo Rose "2," and his trip from San Diego was a total waste.

WORLD WAR II

By late 1940, the U.S. Selective Service was starting to bear down on young Americans to increase the size of our standing Armed Services in anticipation of the war we all knew was coming. My Selective Service status was uncertain. I knew that if I was drafted, I would have no choice of service or job. There were a lot of people in my position. We knew the world was becomingly increasingly unsafe for freedom, and we were not only willing but eager to do whatever it took to make it safe again. But if our country was to go to war, we wanted to be where we could be of the most use, and we were reluctant to allow ourselves to just be drafted and shipped somewhere to do something we weren't really suited for. So I decided to enlist. First, however, I took advantage of an opportunity to go up the Colorado River by boat with an All-American Canal surveying party, a venture I described in Volume 1. Details of that trip have always stuck indelibly in my mind. After all, for all I knew, I would never see the river again.

After the river trip, I left my 1938 Chrysler coupe, which by that time had 20 or 30 thousand miles on it, with my parents in Yuma. I gathered up $20 in cash and caught a bus to San Diego. On January 11, 1941, I went to Fort Rosecrans and enlisted for a three-year stint in the U.S. Army. The unit I was assigned to was the 6th Regiment of the Coast Artillery Corps (CAC), a Regular Army regiment assigned to the harbor defenses of San Francisco. It was a big day for me; they gave me the serial number 19044261 RA, and I've never forgotten it. The "9" is for the 9th Corps Area of the United States. (Number "1" was on the East Coast, and the numbers increased to the west, so that Number "9" was in the westernmost area.)

San Francisco

So there I was in San Diego, an Army man. In a few days they shipped me to San Francisco by rail, and I reported at Regimental Headquarters at Fort Winfield Scott, a part of the Presidio, at the south end of the Golden Gate Bridge. There were five or six three-story barracks around the parade ground that were well kept—spic and span—making a good first impression. Because it was winter, the first thing they did was to issue me my "woolens." Actually, the year-round low temperature in San Francisco is about 56 degrees, the same as the ocean. The coastal fog maintains the warmth—such as it is—in the mornings. In the afternoon, if the fog burns off, the thermometer can go into the 70s, but as soon as the fog rolls in again, there you are, back at 56. So we never really needed "cottons." Within a week after signing up I learned my General Orders,[1] was issued my

1. General Orders were principally fixed instructions for Army sentries.

1903 Springfield rifle, and, after firing a few rounds on the range, was walking sentry post. It is hard for civilians to appreciate the importance of a rifle to a soldier. It was perhaps our most important military possession and a great responsibility. To this day I can rattle off my rifle's serial number—822332.

Just about the time I arrived in San Francisco, a detail had been sent to Chicago to bring back a large group of the first draftees. To make room for the new recruits, my unit, Battery B of the CAC 6th Regiment, moved 4 or 5 miles down the coast to Fort Funston. This name had a special significance for me. The fort was named for the general whom I had watched reviewing troops in Yuma on an alarmingly white steed about 25 years earlier (see Volume 1). At the fort, we holed up in old CCC barracks near the beach. There were about 15 of us RA "recruits," all of whom had been assigned to duty without receiving any basic training. In fact, we helped train other recruits. For the most part, we just walked post and got an occasional morsel of instruction. In contrast, incoming draftees got the full treatment. Our unit was captained by Richard Moorman, a West Pointer, class of '32.[2] There was one West Point lieutenant and two other officers, who were young reservists. The first sergeant retired soon after my arrival, and John Zarko took over. Zarko was a three-striper when I first knew him. He had seven hash marks, indicating 21-plus years of service. Actually, according to the old-timers, if he could have counted his "bad time"[3] he would have already had his 30 years in. So he really knew the ropes. He was *good*; he was right on the job. He was THE BOSS!

At our new post at Fort Funston we were right opposite Fleishacker Zoo, about a half mile from the end of the Taraval Street streetcar line. The amenities of San Francisco were not brand new to me. I knew the zoo and the cable lines and a lot of the streets from the summer of 1923, when Mother and I had been there visiting my father; he was in the Southern Pacific Hospital, at the east end of Golden Gate Park, recovering from his second broken leg. We were there for six weeks, living in a rooming house at 1640 Hayes Street. Market Street was the

---

2. Moorman retired as a colonel. His first cousin, Robert Moorman, also class of '32, was commander of Fort Huachuca circa 1965; he was then a two-star general. He and his wife called on us in Patagonia in 1970 or so.

3. It didn't take much to get "bad time" in the army. One little thing and there you were. I know! They gave me a little "bad time" once myself! Great militaries have to do things like that. "Bad time" is also known as "lost time." When an enlisted man's discharge time comes due, he must serve an additional day for each day he was in the guardhouse.

main drag then and still is today. It was 100 feet wide, and it seemed that there were always trolley cars going by. There were four trolley lines—two each way.

When I was stationed in San Francisco, girlfriends showed up from my days in Morenci, Bisbee, and Tucson. It was good to see friends of any kind, but, of course, the girls were especially welcome! San Francisco was a good place for dates because there was a lot to do. First-run movies at Market Street theaters cost 75 cents for most people but only 35 cents for us Army guys, as long as we were in uniform.

The Battery B quarters at Fort Funston were solid. The regimental headquarters area was not fancy, but those handsome Fort Winfield Scott barracks had taken a lot of waxing and polishing over the years and were always right out front whenever any of the big brass showed up. We had our own cleaning and tailoring shop and a 35-cent (civilian) barber. Later we even had a little Post Exchange, which we shared with Battery C. The "Cs" manned a pair of 16-inch Navy "rifles" just south of Fort Funston. Civilians might wonder what good rifles would be in coastal defense. Well, these "rifles" were actually 16-inch cannons! They fired projectiles (shells) each of which weighed just over a ton; each shell had a rotating band of copper around the base. Interestingly, these rifles had been salvaged from decommissioned battleships when the 5-5-3 treaty[4] had been passed in 1922.

I thought at the time, there at Fort Funston, that Battery B was the only outfit in the Army that manned both antiaircraft and seacoast weapons. We had two 3-inch (fixed) barbette-mounted AA (antiaircraft) guns,[5] as well as two 12-inch seacoast mortars that fired armor-piercing projectiles. These were smaller than the 16-inch shells—they weighed only half a ton! But I found out later, by going there to serve, that Battery A of the 13th CA Seacoast Artillery, on Santa Rosa Island, near Pensacola, Florida, also had two fixed 3-inch AA guns. At Fort Funston I was a telephone operator on the AA setup and a gun pointer on one of our two mortars. These were challenging jobs—as Army positions went—and I was glad to be able to use a little bit of my engineering knowledge and experience.

As I mentioned on page 38, only a few days after I arrived at Fort Funston I received a telegram from Phil Lynch in Tyrone, New Mexico, finally offering me

---

4. The Washington Naval Conference of 1921-22 resulted in three treaties, which attempted to establish stable relationships among the naval forces of the various world powers. In one of these treaties, focusing on arms limitations, a 5-5-3—1.75—1.75 ratio was established among U.S., British, Japanese, French, and Italian battleships. That is, for every 5 U.S. and British battleships, Japan was allowed 3 and France and Italy 1.75. Maximum total tonnage was limited, as well as specification of a maximum single-ship tonnage of 35,000 tons (Luttwak, 2004).

5. A barbette battery is an elevated site on which guns can be mounted to fire over a parapet. Guns so positioned are referred to as "en barbette."

the lease on the fluorspar that I'd asked for back when I was in Bisbee. Two months after the fluorspar telegram came, my draft board at El Cajon, California, also way behind the times, inquired as to my health and whereabouts. I didn't bother to reply. I had a harbor or two to defend.

In the Army position and rank change rapidly during the first year. Over a nine-month period, I went from recruit, which paid big money—$21 per month—to private first class at $36 per month, to corporal, where I became a rich man ($54 per month). Of course, it never occurred to me or my friends then or now to complain. Why should I have complained? I was defending freedom! Looking back, I do wonder sometimes whether people today appreciate how little money matters as compared with their freedom.

When I was a corporal, I became the first enlisted man from Battery B to attend Officer Candidate School. I always thought that Sergeant Zarko, who liked me, had something to do with this opportunity. I was in the second OCS class at Fort Monroe, Virginia. I went there in October 1941 and was still there when Pearl Harbor was attacked. I was commissioned as a second lieutenant in AA and Seacoast Artillery just before Christmas 1941 and was given two weeks of leave, which I spent in Yuma and Bisbee, traveling by rail.

Fort Bliss

A few of us from my OCS class had been assigned to the 79th CA (AA) at Fort Bliss, at El Paso, Texas. The selection process was not mysterious: they did it alphabetically! It was Kramer, Lenon, Linke, a Mc-something that I forget (other than that it wasn't MacLennan), and one or two more. Bob Linke and I became good friends. He was a civil engineer, about my age; his officer serial number was within a dozen or so of mine. Our battery was pretty good, considering that the officers were green. We weren't all officers. Our first sergeant was a draftee who had been promoted directly from battery clerk, in which capacity he had been a corporal.

We were each assigned a small winterized tent at Logan Heights for the times we were on duty overnight, but we could live off post if we wished. The tents were 10 by 10 feet and were pretty nice as tents go: each had a cot, a table, and two folding chairs. Latrines and showers were on down the line. A few men just stayed in the tents, but most of us, including me, chose to live off base. A good buddy of mine, Rube Lankford, a World War I sergeant major whom I knew from my Bisbee days, was working in the El Paso Post Office at that time, and he and his wife found me a room for rent with one of their neighbors. Rube was a veteran of World War I and, as a sergeant-major, gave me a lot of good advice on how to get ahead in the Army. He had served in France and had married a French girl. They had come back to New York with the very last shipload of World War I troops, in 1921.

Our battery and about half of another were on steady guard duty at the Army Air Corps' Biggs Field. We were assigned for a week or 10 days for each hitch, so that's how I spent most of my on-duty time. You might well ask why a Coast Artillery man was at an inland "port" on the Rio Grande—which, despite being called "grande," was narrow even then—helping to man a series of antiaircraft weapons that protected a U.S. Air Corps field practically on the U.S.–Mexican border. Well, the 79th was getting ready to ship out for Europe.

When the unit departed in the spring of 1942, however, Bob Linke and I were left behind! We were assigned to the Reception Center, where we processed draftees and enlistees from Arizona, New Mexico, and west Texas until July 1, 1942. We knew that it takes draftees to fight a war and that somebody has to process them, but both of us were crushed to have been left behind when our unit went to Europe. The Army had prepared us to fight, and that's what we wanted to do. But there we were—processing recruits. They thought we were too old: Linke was 37, and I was 34. The recruits poured through our center by the hundreds, and it took a good-sized unit to process them. We were assisted by 15 or 20 Japanese American (Nisei) soldiers—fine soldiers and good fellows—who had been pulled out, one by one, from regiments that were headed for the Pacific Front.

Going to Pensacola

Linke and I got along very well. We outfitted recruits, including many I had known previously in Arizona and New Mexico as civilians and even a few I had gone to college with. We did all right, except that the assignment was, for lack of a better word, boring. I got so fed up with the monotony that I even wrote to the commanding general and offered to resign my commission and reenlist if that would help me get overseas. I can't imagine that the Command was pleased that I wanted to quit my recruiting duties. Nevertheless, I have no reason to suppose that my offer to become a buck private again was connected with my getting an order to report to the harbor defenses of Pensacola, inasmuch as Linke also received such an order.

Linke had a car, a 1938 Buick coupe, which we drove to Pensacola in. It took us the better part of a week to get there because government restrictions[6] limited us—and everyone else—to 35 mph! We did stop off briefly in New Iberia, Louisiana, where I wanted to visit the underground parts of a salt mine. We were told that we

6. During World War II, the War Production Board took strong measures to conserve scarce materials. The production of civilian automobiles was suspended, tires and gasoline were rationed, and a 35-mile-per-hour speed limit was imposed (U.S. Federal Highway Administration, 2004).

had no business there, that it was restricted and so forth, but I pulled out my AIME (American Institute of Mining Engineers) card and talked to the chief engineer. While I was talking to him, I noticed a copy of Peele's *Mining Engineer's Handbook* on his bookshelf. I pulled it down and showed him the two articles I had written back in 1937 on large-scale underground mining in the Campbell (copper) Mine in Bisbee. Then we got a blue-plate special tour! It was a clean and efficient mine, and I will always remember seeing 400-to-500-foot-thick chunks of rock salt!

Defending Against German Torpedoes

The degree to which the United States was under attack by German submarines along the East Coast (and by Japanese submarines on the West Coast) was not publicized during the war, and even people who remember the war and the big battles in Europe and in the Pacific often are unaware that there *was* such a threat. But the threat was very real, and our unit was stationed on the Gulf Coast to assist in coastal defense. The Command was mainly concerned about the ship-building towns along the Gulf Coast—Panama City, Pascagoula, Biloxi, Gulfport, and New Orleans. These and military installations such as Fort Barrancas (at Pensacola) and Fort Morgan, which was responsible for the harbor defenses of Mobile, were the most likely targets for enemy submarines.[7]

We reported to the adjutant at Fort Barrancas at the beginning of July 1942 and were assigned to the 13th Coast Artillery at Fort Pickens on Santa Rosa Island, a barrier island[8] a half mile offshore. From Fort Pickens we went by 3-foot-gauge Army railroad[9] to our emplacements. My emplacement was a "bat-

---

7. The 13th Coast Artillery manned Fort Barrancas from post–Civil War times until 1947. Historically, this regiment was responsible for coastal defense from Panama City, Florida, to Galveston, Texas. Fort Pickens and Fort McRae were subposts. New weapons developed during World War II finally made all coast artillery obsolete for the United States, and in 1946 the Army turned the area over to the Navy. Fort Barrancas became a part of Gulf Islands National Seashore in 1971. Following 18 months of restoration, the fort was reopened in 1980 by the National Park Service.
8. This was the site of the prison where the famous Chiricahua Apache warrior Geronimo languished in the 1880s. Geronimo led his people's defense of their homeland against the U.S. military after the retirement of Cochise, another famous Apache warrior. Geronimo surrendered in 1886, spent several years as a prisoner at Fort Pickens, and finally was moved to Fort Sill in the Oklahoma Territory in 1894. Before he died there in 1909, he dictated his autobiography (Geronimo et al., 1996).

tery" that consisted of a single 155-mm cannon. As in San Francisco, the harbor armaments were scrounged from a variety of seemingly unlikely places. Our lone gun, a French-designed 155-mm GPF (known both as *Grande Portee Fillioux* [Long-Range Fillioux] and as *Grande Puissance Fillioux* [loosely translated: "Fillioux's great power"]), bore the serial number 32 or so. That meant that it was only the 32nd cannon of that model produced. It had been used in World War I. The gun had hard rubber tires, and its gun-record book stated that it had reposed in the Smithsonian Institution in Washington as an exhibit from 1920 or so to the time of Pearl Harbor.[10] If anybody wondered whether our military was serious about defending the homeland, the idea of emptying the museums of cannons is a simple statement!

Weather-wise, Fort Pickens was an excellent post. It was a little warm and muggy on occasion, but I don't remember too many uncomfortable days. At one time or another I was stationed at most of the other emplacements along the Eastern Gulf subsector, including Panama City, Florida, and Fort Morgan, Alabama, the place where the famous saying "Damn the torpedoes, full speed

---

9. Narrow-gauge railroads in the United States were built largely with a 3-foot rail-to-rail spacing. These railroads were much less expensive than standard gauge (4 feet 8½ inches) and hence were more appropriate for military purposes, in which loads—no matter how strategically important—were moderate. In World War II, the U.S. military expanded its railroad system whenever and wherever it was strategically necessary. Because narrow-gauge railroads were becoming economically unprofitable in industrial operations, the military generally had little trouble purchasing needed equipment. However, in Alaska, the possible site of a Japanese invasion across the Bering Straits, defense requirements were so great that the U.S. Government took over one narrow-gauge railroad, the White Pass and Yukon. The military operated the railroad until it became clear that the threat of a northwestern Japanese invasion was virtually nonexistent (White Pass and Yukon Railroad, 2004). Most narrow-gauge railroads were "feeder lines" that connected to standard-gauge lines. Economically, the narrow-gauge lines became economic liabilities, as the cost of transferring loads from narrow- to standard-gauge equipment increased (Hilton, 1994).
10. The GPF guns, designed in 1917 by a Frenchman named Fillioux and later adopted by the U.S. Army (Anonymous, 2005), were used in coastal defense. One of these guns, mounted on a mobile carriage, is displayed today at Fort Sill, Lawton, Oklahoma. These guns were often emplaced in sets of four on circular, or semicircular concrete "Panama mounts" (American Forts Network, 2002).

ahead" was uttered.[11] I was also stationed at one time or another at Pascagoula and shipyard towns like Biloxi, Mississippi, and Port Eads and Burrwood, Louisiana. The latter emplacements are located at principal outlets of the Mississippi Delta, some 130 miles below New Orleans. I laid out the coordinate systems for all of these setups so the guns could fire at unseen moving targets. As a "reward" for my useful work I got to ride all of the tugboats towing the moving targets during firing practice. This quaint Army custom ensured that all of the firing data that had been furnished to the gun crews were accurate, inasmuch as tugboat passengers might be blown off the ocean if the information was wrong! As an engineer, I still find the process we used to be fascinating. Readers who would like to know more about how we could hit moving unseen targets will find some of the details in Appendix III.

As important a port as Pensacola was in the war effort, the Mississippi Delta and Mobile Bay were even more vital. Many of the coastal emplacements protected shipyards because they were inviting targets for German submarines. The subs sometimes lay just out of range of our 155's, waiting to attack ships that were either unarmed or, if armed, were not yet fully manned. They knew what we had, and we knew what they had, and we knew that they knew, and they knew that we knew. So it was just a question of who was better at playing the game. Weekly reports (top secret, of course) at headquarters indicated that there were 20 to 25 German subs active in the seas in the Gulf region. They were refueled by small vessels disguised as fishing boats or by boats flying neutral flags in equally neutral Caribbean ports. Mostly the subs—as befits their name—were out of sight. But we did see enemy subs several times. I remember being told by fellow officers that a German submarine had aimed a torpedo at a ship emerging from the main channel of the Mississippi. It missed, but it got part of the jetty alongside one of our gun sites as a booby prize! I don't think the German subs ever hit any of the ships in sight of our battery setups. I suspect that they must have hit something somewhere, but, as far as I know, it was not reported in the press.

---

11. In August 1864, Rear Adm. David Glasgow ("Old Salamander") Farragut, who had achieved fame and rank by opening the waters of the Mississippi Delta two years earlier, commanded a Union fleet in the Battle of Mobile Bay. Farragut, a 60-year-old southerner who had remained loyal to the Union, was aware that the bay was heavily mined. He nevertheless ordered his fleet to charge the enemy's positions with the famous words "Damn the torpedoes! Full speed ahead." The "torpedoes" in Civil War battles were the mines that menaced his fleet, not the more familiar naval munitions of the 20th century (Patrick, 2004).

Each site presented unique problems. Laying out baselines and doing the required triangulation at the gun emplacements located at the mouth of the Mississippi was one of our more difficult challenges. The "land" at the sites at Burrwood and Port Eads was barely that. For example, if you jumped up and down a few times on the parade ground at Port Eads a growing circle of watery silt would come up around you, like quicksand. Before I got there, a bulldozer left sitting overnight on the ground had sunk out of sight—forever!

Often, at either of the lower delta sites, if I tried to wade ashore from a rowboat carrying a 20-plus-pound transit, I'd sink into the muck virtually shoulder deep. With the transit on my shoulder, I temporarily weighed 175 pounds! One of the ways I coped with this challenge was to walk on my knees to shore; the same amount of my body would remain above the surface. But when I was on my knees, I had a broader base and didn't sink into the mud so far. Another method that we used when walking across shallow and narrow drainage ways in the marsh was to wait until low tide and then look for a spot with a small piece of driftwood near the center of the ditch. We would then use that driftwood as a sort of stepping stone. One of my master sergeants making that maneuver discovered during such a hop that his "log" was actually the back of a half-grown alligator! Like the log, it was on and off—fast! For that application, the alligator worked just as well as a log would have, but it was more exciting!

While I was stationed at Pensacola, I rewrote the Fort Record Book—essentially from scratch. This was a lengthy document that contained the history and technical details of the fort, including a massive loose-leaf leather-bound "archive" containing the data required to fire all the guns. It was highly classified at the time. Even now I can't describe the document fully, except to note that each page that I had signed was initialed by someone in the Master Gunners Section.[12] All of our reputations rested on getting it right! For this work I got a letter of commendation and a rating of superior. I should have been promoted, but there wasn't a vacancy. In our unit promotions were strictly by time-in-grade, rather than merit.[13] There was one exception. Our chief was promoted all the way up from first lieutenant (reserves) to lieutenant colonel. He had been an elevator operator before he entered the "reserve" military. I guess he knew how to move up in life! He was "in" with the management and next in line every time a vacancy came up.[14]

---

12. That was a long time ago, of course, but the declassification process sometimes doesn't move very fast. It is not feasible to declassify all of the millions of documents that have been classified over the years.
13. Except for West Pointers, who were, quite properly, promoted on merit.
14. It was experiences like this that convinced me that I should be my own boss as soon as I could arrange it.

Finally, in about June 1943, after 18 months as a second lieutenant, I was promoted to first lieutenant, with a pay boost to $175 or so per month. Bob Linke shared my fate with respect to promotion. It seemed at that time that our alphabetical "brotherhood" would continue for the whole war! As it turned out, we *were* together for well over two years.[15]

Corps of Engineers

By early 1944, after I had been at Pensacola for a while, it became increasingly clear that a little occasional mischief from the odd sub was all that would ever happen on the Gulf Coast. Therefore, if I was expecting adventure I would have to go elsewhere. About that time, a call came through for officers with engineering degrees in the CAC to volunteer for transfer to the Signal Corps or Corps of Engineers. Bob Linke wasn't interested, but I was, and away I went to Fort Belvoir, Virginia, for instruction in building Bailey Bridges[16] and other military structures. We also learned, if worse came to worst, how to do demolition work.

In April 1944, upon completion of two months of training, I was assigned to the 1318th Engineer (General Service) Regiment, at Camp Sutton near Charlotte, North Carolina.[17] We spent a lot of time in the Carolinas, building timber highway bridges for North Carolina counties and doing other miscellaneous assignments. The rank and file of our regiment was African-American, and about one-third of the officers were, too. This contrasted with the management of many segregated regiments, in which most or all officers were white. In those days, of course, many Army units were segregated. Higher-ups in the command

---

15. After the war, Linke became a major in the Reserves, moved to Benson, and then moved again to Scottsdale. I saw him from time to time. He died in about 1956.
16. The Bailey Bridge was one of many technical innovations that came directly from World War II. Sir Donald Bailey, an Englishman, designed this structure in 1940, to accommodate an explicit requirement for a 40-ton-capacity bridge. Production began in July 1941, and by December of that year the Bailey Bridge was in use in combat (Mabey and Johnson, 2005). The relative ease with which the bridges could be assembled contributed greatly to Allied logistical capacities and ultimately to victory.
17. Regiments, which are usually commanded by colonels, are primarily responsible for synchronizing the plans and actions of their subordinate units—usually two or more battalions—to accomplish a single task for the division or corps of which they are a part. Corps are the Army's largest tactical units, tailored for the theater and mission for which they are deployed.

wishing to keep one jump ahead of possible critics and to avoid charges of discrimination against minorities, assigned only officers with ratings of "excellent" or better to such units. However, I never knew an officer to be reassigned away from an African-American unit, no matter what he did or how low his ranking fell. For example, a second lieutenant in another unit attacked a Black staff sergeant with a pick handle, but all he got was a reprimand and a transfer into our regiment. So the treatment of African Americans then—even those serving their country well—was quite different from the way it is today.

I Knew Where We Were

Finally, I got my chance to take part in the actual war! Our regiment was ordered to serve in the European sector. We were sent first to Camp Shanks, New Jersey, and shortly thereafter, on October 31, 1944, we shipped out of New York Harbor on the British troop ship Moreton Bay. Our ship was part of a huge convoy. No shipboard announcements of our position were ever made, and I never ran into anybody who knew for sure where we were going. But I had my trusty Brunton[18] along and took solar and stellar observations of position twice daily. I was essentially using the Brunton as a sextant. I shot the sun at noon and the North Star at dusk.[19] Longitude was something of a problem for me, but I overheard (British) civilian deck hands grousing about "these damned 9-knot convoys." Their complaint was that the convoys were held to that speed because of the presence of a large number of Liberty Ships[20] that could go no faster. So every day I just plotted the distance on a big map of the Atlantic, along with my version of the latitude, and I had a reasonable estimate of our position. I was perfectly open about what I was doing, and no one ever objected.

---

18. The Brunton is a civilian "pocket" compass with a dial that can be set to make its sights read "True North." It also has an adjustable level bubble that reads vertical angles to the nearest degree. I still have mine, and even today it works just fine. It is constructed of a very heavy square aluminum frame with a magnetized compass needle under glass.
19. Shooting the sun (or moon or a star) involves aiming a sextant to measure the angle between the horizon and the celestial body. With that, an accurate watch, and a current-year Nautical Almanac, the latitude and longitude can be determined (Holladay, 2002).
20. Liberty Ships were cargo ships made famous by that name during the war. They were turned out by the hundreds in Gulf shipyards. Protection of those shipyards was perhaps the most important duty of our Coastal Defense unit.

We stayed away from the "Great Circle"[21] shipping lanes, sailing virtually due east at the latitude of New York City until just short of the Azores and then shifting to a northeasterly course. I knew within 50 miles where we were at all times. I knew when we would sight land and what lighthouse island (Bishop and Clerks) it would be. Twelve days after we had embarked from New York City, we entered the Irish Sea and docked at Liverpool on November 11, 1944. We settled into British Army quarters at Maghull (pronounced McGull), which was situated near the end of a trolley line from Liverpool.

A Full Regiment of Engineers

Because there are virtually no single jobs large enough to keep a full regiment of engineers in work, Company B stayed at Maghull, and the rest of the regiment split up into smaller units that headed south to winterize hospital tents in Cheshire and Wales. As soon as a group finished one "dispersed hospital," it would leapfrog other groups that were still working and would begin winterizing another unit farther south. When the winterization mission was nearly completed, all of Company B headed south, via London, by rail, and set up headquarters near Shaftesbury. This served as a base to reunite the men of our unit and prepare us to ship out to France.

My platoon was assigned to a U.S. Army hospital at Grimsditch, south of Salisbury. We lived in winterized tents. At the hospital we built covered corridors through which patients could be wheeled in gurneys between dispersed ward Quonsets[22] and a much larger Quonset that was used as an operating room. It is very difficult to do sturdy brick or stone work at temperatures at which mortar freezes. Fortunately, this particular winter happened to be mild, and we were able to lay brick for 24 days during January. We worked a seven-day week and knocked off for only about three days or so during heavy snow at the time of the Battle of the Bulge on the mainland.

---

21. The Great Circle of a sphere is a circle, the plane of which passes through the center of the sphere. It can also be defined as the segment of such a circle representing the shortest distance between two points. It can best be illustrated by stretching a rubber band between two points on a globe—its position will mark the shortest route connecting the two points.
22. Quonset is a trademark for a prefabricated structure made of heavy corrugated sheet iron. Quonsets were built so that their semicircular roofs curved downward to form walls; they were often used to house military personnel and materiel.

When we weren't working on hospitals, we were pulling up landing mats. These mats were made of interlocking sheets of pierced sheet-steel. Once they were put in place, weeds and grass grew through them and held them down. There were quite a few of these mats in the south of England. They were used as takeoff fields for glider invasions.[23] When they were taken up they were baled into bundles and shipped to France, where they were reassembled. Civilians may not appreciate how mobile large structures can be if they are part of a war effort. In February, I was made Battalion Adjutant. As such, I was sort of a construction superintendent and had the pleasure of keeping all the jobs going, and that was its own reward.

Shipping Out to France

In early 1945, we all shipped out to France from Southampton. I was one of about 100 troops sharing the Liberty Ship Samuel Colt with all of our heavy equipment and trucks. Most of the troops went into port near Le Havre on another ship, but we went up the Seine and unloaded at Rouen, with the assistance of civilian stevedores, who were Algerian, if I remember correctly. Then we all went up virtually to the front, at Bruyeres, a picturesque village just south of Laon in the Department of Aisne.

In and near Laon, we repaired and maintained a supply depot and sawmill and, with all the help we could extract from a thousand or more well-scattered German prisoners, did other types of jobs, including highway and railroad repairs. In addition, again using German prisoners whenever possible, we ran several portable sawmills in the Forest of St. Gobain. One segment of this forest was the area from which the Germans had fired Big Bertha[24] on Paris during World War I. There had been other guns, as well. Many of the older standing trees that we were harvesting had big chunks of shell fragments from that long-gone war imbedded in them! These fragments played havoc with our circular saw teeth, of course. When we left Laon I was Commander of Company F, the unit I had built a county highway bridge with in North Carolina (see Appendix IV).

---

23. Gliders, piloted by infantry sergeants, were loaded with armed "invaders" lined up in ranks, each with a big hook on top. These hooks were "lassoed" by slowly flying tow planes and were dropped off at assigned spots to "coast" to battlefield positions, land, and unload into battle.
24. Big Bertha was a type of long-range rail-mounted cannon the Germans used in World War I. It was named for Bertha Krupp, daughter of armaments manufacturer Alfred Krupp. It was fired from various railroad spurs about 60 miles north of Paris (Miller, 1930).

Laon is a major railroad city with big marshalling yards, and in the late stages of the Battle of the Rhineland (February-March 1945) we repaired a lot of bomb damage on railroad lines. However, I was only at the actual front once. In that adventure, we were trying to pick up a hundred or so liberated Polish POWs.[25] The Allied lines had moved forward rapidly, and the Poles had been left behind. We were given the assignment because we had the biggest expendable pool of trucks and drivers and I was one of the few officers with considerable map experience. So they gave me a map of the area in which we were to search. I still have that map. I never told my two-man truck crews just how close we were to the front line. We looked hard for those POWs. We talked to a lot of people in the little towns we passed through. Many of the village folk were up and about even at 2 or 3 a.m. In each town we saw streaks of candlelight in the windows. Many of the local people knew about the POWs, but I had been given a mangled town name, and I am unsure to this day that we were in the right place.

Finally, we ran out of road. I took one man from each truck and went ahead on foot. We had gone only a short distance when we were challenged. I thought at the time that the language was Russian. But when you're challenged, it doesn't really matter what the language is—you know what they are saying! It was a very dark night, and, as it turned out, I was in peril. I didn't realize it until later, but by then I was all alone. My higher ups had given me truck drivers from the regimental motor pool. They were not part of my regular unit, so, since they were not accountable to me, they had simply melted away. Furthermore, I had outrun the map that was given me, so I didn't know exactly where I was. As I remember, there were two towns near Laon, perhaps 3 kilometers apart, with very similar names—something like Cerny-les-Bucy and Bucy-les-Cerny—so I may have been in one of them. Anyway, because I thought I was backed up by at least a dozen "night fighters" I was talking pretty sassy to the sentry, in French, mainly, but trying a little German, too. I finally got the dope on his commanding officer, who turned out to be an American buck private paratrooper who had gathered up a little fighting force of his own. Since that was none of my business, I said to my men, "Let's go." But, to my great surprise, no one was there except me. All I had was my Garand (M-1) rifle.[26] I don't suppose the sentry ever knew I had had anyone with me in the first place. The wonder is, really, that

25. This may not sound like a job for an engineer, but my platoon was part of a General Services regiment. We also supervised a group of 300 to 400 German POWs. They had to be guarded around the clock, and when we began that job, they were not enclosed. We subsequently built a double fence around them to make it easier to guard them. The POWs, under our supervision, sank a 30- to 35-foot well in the camp (and timbered it, mining fashion, using oak timber we had felled), in accordance with requirements of the Geneva Convention.
26. I normally carried a 30-caliber carbine, but I considered it to be a bit puny for front-line duty.

he and I didn't shoot each other or that I didn't get shot by my own men. Fortunately, they had all returned to our trucks. I also returned to the trucks and drove back to headquarters empty handed.

Off to Marseilles, Slowly

I was in Laon on VE Day. The day the peace treaty was signed—May 5, 1945—I had gone to Chateau-Thierry (of World War I fame) via Soissons to rustle used railroad ties from a supply depot to build a railroad spur to one of the sawmills we ran. Late that afternoon, on our return to Bruyeres, via Rheims, we saw vehicles full of generals going to the peace conference. As I said above, I had just been named Commander of Company F, and in that capacity I was handed a fistful of maps, preparatory to leading a half-mile-long convoy to Marseilles a few days later. First, however, we were asked to head up the U.S. Army's VE Day Parade, in the walled-in part of old Laon. So the Battalion Commander and I, as his adjutant, followed by the rest of Company F, paraded afoot. Remarkably, in the next day or so, on the streets of Laon, a French lad about 10 years old flashed a photo postcard at me, depicting me and my company, marching in full parade trim. He was pointing to my picture and saying "voo, voo" (*vous, vous*; "you, you"). Sure enough, 'twas I! So I found a little store that was selling the cards and bought a few to mail home. I still have one.

VE Day parade, Laon, France, 1945. Author third from left. From a postcard placed on sale shortly after the parade. Photographer unknown.

Several days later, we started off, in convoy, on a three-day trip—at 35 miles an hour—to Marseilles, whence, rumor had it, we were destined to ship out for Okinawa. I had the maps and was given an Army command car and driver, and away we went. The convoy was ordered to move slowly, at the same speed as the civilian traffic, to save gasoline and tires. We were also ordered to take roadside breaks for the last 10 minutes of every hour.

Visiting a World War I war bride's French relatives in Troyes, France, a few weeks before VE Day, 1945. Mme. LaRuelle and her daughters, Claudette and Viviane, are behind the author, who is sitting on the bumper. Photograph by M. LaRuelle.

The pace of the convoy was so slow that I found a way to put it to use. I turned over all my maps to one of my Company F lieutenants, John Sankey,[27] and put him in the lead truck of the convoy, just behind my command car, after marking out the route they were to take. Then, upon entering Troyes, I peeled off and, with my driver, paid a visit to the LaRuelle home. I had come to know

27. John was a mighty fine gent from Richmond, Virginia. He was the only member of the regiment that I looked up after the war. He died four or five years ago.

the LaRuelles through Rube Lankford—my friend from Bisbee and Fort Bliss. Rube's wife had told me that she had a married relative still living "80 miles out of Paris." It turned out that they were at Troyes, an ancient silver-working city. So I had gone earlier to visit them before our convoy shipped out.

We couldn't stay long, but it was close enough to lunchtime that one of the girls rushed out and picked an apronful of "petti-pwa" (*petits pois*)—delicious tiny green peas. My driver and I had a fine meal and a good visit before having to drive off. We still hear from Claudette Renault, the older LaRuelle daughter, and visited her twice in later days as civilians, on Utah Beach and in Rouen. Claudette taught school in England for several years and then moved back to Rouen. Her French husband died young and she raised three fine sons alone. Claudette's younger sister, Viviane, married a young Norman lad, but he, too, died young, only a few years after they were married. She lives on Utah Beach.[28]

After leaving Troyes, by traveling at close to 37 miles an hour and not taking the hourly 10-minute breaks, we caught up with our convoy by midafternoon and took our assigned place in the lead of that portion of the regiment. After I rejoined the convoy, we went through Lyon and then, two days later, Avignon. We "reported in" near the ancient Roman city of Arles.[29] There a regiment of engineers was setting up tents and prefab plywood buildings to accommodate us and dozens of regiments coming in while we all awaited the arrival of a troop ship headed for Japan. We thought then that we would be going east through the Suez Canal.

---

28. Utah Beach is the westernmost of the five beaches where the D-Day landings during the Battle of Normandy took place on June 6, 1944.
29. Arles is home to a Roman amphitheater, Les Arènes, that was built around 90 A.D. and can still hold 20,000 spectators. Gladiator fights and chariot races were held there until the end of the 5th century. During the Middle Ages it was a fortress. Today it is the scene of cultural events and bullfights. Les Arènes has been listed as a French national monument since 1840 (France Travel Guide, 2005).

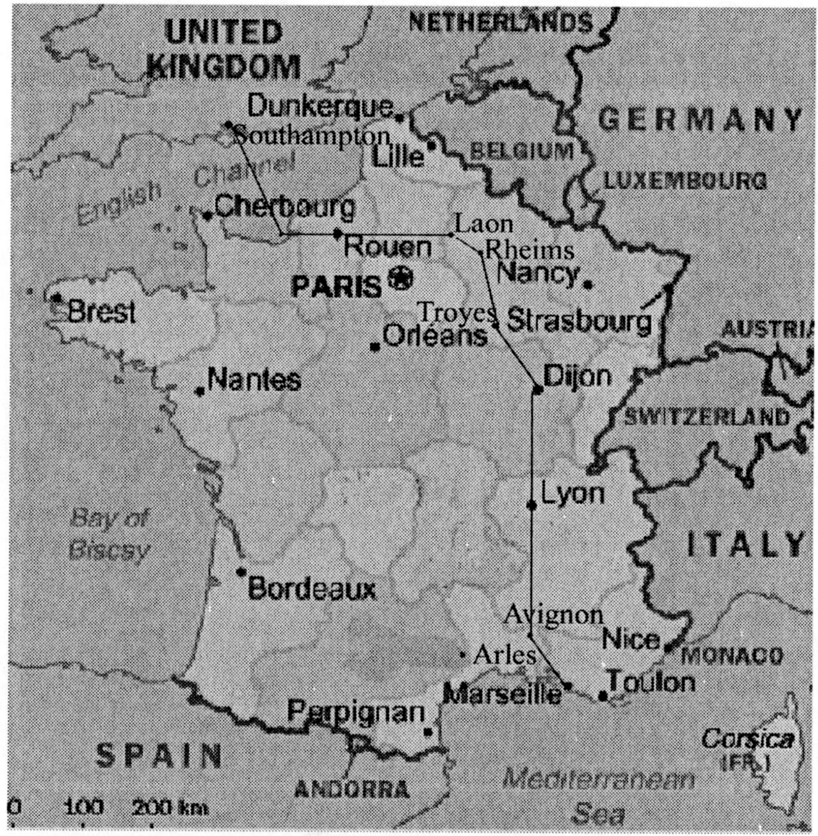

My "vacation route" in France.

Marseilles Was a Rough Place

Once in Marseilles, we thought we would be shipped out almost immediately. Our supply sergeant was a good hard worker and reoutfitted all hands as required. He did such a good job that within two weeks after we got to Marseilles, he had almost worked himself out of a job. I suggested one afternoon that he could leave things with one of his assistants and go "downtown" to see the sights. Then I went on about my business. Late in the afternoon, I happened to see the sergeant back in his supply room. So I said, "Sergeant, you said you wanted to see the city in daylight; it's going to be too late soon." He replied, "Oh, Lieutenant, Sir, another sergeant and I already went. Do you know where that little round-about park is, with a statue out in the center?" "Yes, sure, Sergeant." "Well, there's a big bar right there, and we got off the bus and went in there. We were drinking our beer when two tough-looking Frenchmen got in a fight. So here came the gendarmes. They held each man up by one arm, like so [demonstrating a hammerlock with one hand directly behind his neck]. And do you know, the meanest looking Frenchman reached inside his shirt with his free hand and pulled out a little dagger and killed the other guy with one stab—with the gendarmes still holding them both in hammerlocks! So, I said, 'Sergeant Jones, let's get out of here and go back to post. I want to be able to get back home and see little Junior!'" They both left half their beers behind and came back "home," where it was safe. They were glad to have survived their trip into town! Some of our troops were killed in brawls in Marseilles. Some nights 10 to 15 men would be killed. I have never seen a tougher place!

Company F officers on convoy, France, 1945. Author at far left. Photograph probably taken by our driver.

Tough as Marseilles was, I sometimes went to town myself—in the daytime, of course—and one afternoon at the Marseilles Zoo, I overheard a white sergeant from Texas talking to a couple of young lads in what I recognized as Spanish. When he left to go back to his duties, I went over to talk to them, using the "border Spanish" I had learned back in Arizona. It turned out that the lads were from Spain but had left that country because they were anti-Franco and had picked up jobs in Marseilles. One of them told me that certain streets in Marseilles were so tough that when the Germans occupied the city they didn't dare patrol the area! There was a lot of black market money there, and it was fought over continuously.

Here's an example of the things that happened in the rough-and-tumble black market skirmishes. One day an "entrepreneur" staked a dozen or so local produce peddlers to a snappy-looking little donkey cart apiece, and they all got off to a fast start in business.[30] But less than a week later, at about 4 a.m., every cart blew up! Each had been loaded with a half ton or more of powerful explosive, along with a time bomb. I was told that the explosions completely wiped out five or six full blocks. That may have been a bit of an exaggeration, but the new peddlers certainly got the message!

Marseilles to Okinawa

Finally, in early July, having lived through our Marseilles visit, we headed out for the safety of Okinawa—to our surprise—via the Straits of Gibraltar and the Panama Canal. It would have been closer to go via the Suez Canal, but at that time the Japanese still held Singapore and controlled the nearby waters. Ours was a 20-plus-knot ship—considerably faster than the 9-knot pace with which we had crossed the Atlantic. Nevertheless, it took us 63 days to get to Okinawa, which was, after all, considerably more than halfway around the world. We went unconvoyed, although we did have 5-inch guns, manned by U.S. Navy crews.

I again passed my time by attempting makeshift navigation. My accuracy was still about plus-or-minus 50 miles, as it had been when we crossed the Atlantic earlier. In the Caribbean, however, I found that I wasn't right on the mark when my map showed us on a course that would have bisected Puerto Rico! We did pass close to that island, but since they hadn't put an east-west canal across it, we had to go just north of it through Mono Passage.[31] It was very foggy, but fortunately the ship's navigator had better bearings than I did. I was still pleased, when the fog lifted a little bit, to see Mono Island, which I had identified on my map. A Puerto Rican captain in our regiment identified the island positively. So I wasn't far off. We were 10 days or so west of the Panama Canal when, on August 6, 1945, the atomic bomb was dropped on Hiroshima. Ships just a week or so behind us were turned back, but we went on ahead. We were tied up several weeks each in Eniwetok and Ulithi by a series of mix-ups, but we finally landed on Okinawa.

30. The entrepreneur had been a German soldier during the war.

31. Mono Passage is an ocean channel between Puerto Rico and the Dominican Republic.

We holed up on Okinawa for about four months, doing work on the air strips at Yontan and Kadena and construction at a U.S. Army general hospital. We did all the big jobs there and some hurry-up ones, too. I was actually memorialized on the island in a very small way. I heard later that for some time my company's street near Yontan was represented on some maps as "Lenon Lane."

Going Home

With many others, I finally left the regiment and got out "on points." In mid-December 1945 I shipped out for Seattle on the Navy transport St. Mary's. On the way home we had a fine Christmas dinner aboard ship. That night we crossed the International Date Line,[32] so the next day was Christmas all over again. By then I knew the mess officer fairly well, but although I made what I thought was quite a convincing argument for a second feast day, my views did not prevail.

When we were but a mere two or three days west of San Francisco, headed for Seattle, the main-bearing of our ship's drive shaft failed, and we could barely sail fast enough to avoid drifting away. As if that weren't bad enough, there was a fierce storm heading our way; we needed a tow urgently, so we sent out an SOS radio call. Fortunately, the call was picked up by the USS Nashville, which was cruising several hours to the south of us. The rescue was a remarkable one. The Nashville's Chief Bos'n's Mate came on the deck of his ship with a rifle-like "line gun" in hand. He fired us a cotton line of clothesline weight, which our own Bos'n reeled in. It was then a matter of tying on successively heavier lines. We wound up being towed east for several days, by our own anchor chain, directly to the Golden Gate. Most of the way we were passing through the storm. It was fierce! It rearranged a good many things aboard both of the crafts involved. We were close enough to shore that we could hear radio broadcasts from various news services, all relating the news of our plight as it developed. We didn't know if we were as bad off as the broadcasts made it sound, but we couldn't help but be somewhat alarmed—as my parents in Omaha certainly were!

We finally arrived in San Francisco early in January 1946, and I was mustered out.

---

32. The International Date Line is an internationally agreed upon line running north-south roughly along the 180 degree meridian of longitude. To the east of the line the date is one day earlier than to the west.

PATAGONIA

Patagonia, which has been my home for almost 60 years, is in south-central Arizona on State Route 82—deservedly designated as a scenic highway—about 13 miles southwest of the Sonoita plateau and 18 miles north of the Mexican border. Sonoita Creek, part of which runs year round, meanders through a valley to meet the Santa Cruz River, which is dry except in times of heavy monsoon or winter rains. When Rollin R. Richardson decided that this pleasant valley would be a good place for a town, he selected the current site and arranged for it to become the main town in the valley, replacing Crittenden, which was 3 miles to the northeast. As I described in an earlier chapter, I first saw Patagonia in 1928, during my summer field training at Mowry. Although I spent most of my time out in the hills surveying, I was in town often enough to form a lasting, favorable impression. I would like to share my memories of my student days in Patagonia and also to comment on some of the changes that occurred after my first tenure in town.

When I first saw Patagonia in 1928, the Southern Pacific Railroad's Fairbank-to-Nogales line ran right through the center of town. Southwest of town, the railroad right of way was close to Sonoita Creek. In July 1929 a flood washed out the line at a narrows about 3 miles southwest of town, and the Patagonia-Calabasas section was never rebuilt. However, even with Patagonia as a terminus, the railroad continued for some time to serve the area as the main means of cattle and ore transport. A considerable amount of ore was shipped from the many mines scattered around, not only in the Patagonia Mining District that I had surveyed in my summer field classes but also in the nearby Toltec and Aztec Mining Districts (see maps on page 143). By 1961, however, both cattle and ore transport volume had diminished greatly, and the Fairbank-Patagonia section was also closed. Because I was the son of a "rail" and have always been close to railroading, I wanted to be one of the last passengers to travel that section. In June 1961 I went on a special trip with other members of the Pimeria Alta Historical Society. I made some historical notes for the other passengers on that trip, and I have included them as Appendix V.

PATAGONIA IN THE LATE TWENTIES

My memories of Patagonia as it was when I was doing my field training in 1928 and 1929 have stayed with me all these years. The streets weren't paved and were traveled as much by wandering burros as by automobiles. In some ways, however, the town wasn't all that different from what it is today. The 1930 Census was to give Patagonia about 500 residents,[1] and there are still fewer than 1,000!

---

1. The corresponding figures for Tucson were about 55,000 and Phoenix (including Mesa) 173,000.

"McKeown Avenue" in the 1920s before Patagonia's streets were paved. Corbett Lumber Company in lower left corner. Photographer unknown. (Courtesy of Bill Bergier.)

One or two trains per day each way stopped at the depot, which I was told dated from 1900. The station agent, a man named Stone, had been a fixture in town for many years. Although the depot no longer accommodates rail passengers, it is still there, hosting town offices. The then unpaved Nogales-Tombstone highway ran down the southeast side of the tracks,[2] along McKeown Avenue, crossed the tracks on Third Avenue, and ran on northeast across a narrow steel bridge that I was told had been salvaged from an abandoned railroad line in Colorado. So in those days McKeown Avenue was the heart of town. Naugle Avenue—which now carries Route 82 traffic—deadended near the site of the present-day waste water treatment plant.

I remember some of the town elders who were active in those years. One was Charles Mapes (1882-1954), who was the longtime SPRR Section Foreman in Patagonia, as his father had been before him. He and his family lived in the two-story section house, which was situated near the place where a gazebo now stands in the town park. People today would never dream that under the gazebo are a

---

2. Patagonia, unlike many western towns, is not laid out on a north-south-east-west grid. The railroad, and the highway beside it, Arizona Route 82, run on a northeast-southwest line. For simplicity in my discussions, even though Route 82 is an east-west highway, I will consider Nogales to be southwest of Patagonia and Sonoita northeast of town.

dozen or more massive concrete pylons that once supported a weighty steel water tank, which supplied water for steam locomotives. In the early 1960s, when Route 82 was being widened through Patagonia, the old railroad section house was moved, intact, to a lot southwest of Cady Hall, where it still serves as a residence.

This home on Duquesne Avenue, which is taller than most Patagonia houses, was once the Southern Pacific Railroad section house in which Charles Mapes and later Pemberton Blake lived. It was moved circa 1960 from its original location between McKeown and Naugle Avenues to its present site. Photograph (2005) by Robert Whitcomb.

Lead had been one of the important metals supplied by the Patagonia mines. However, only a small amount of slag and an SP spur remained to mark where the Empire lead smelter had stood on what was later named Smelter Avenue. A little false-fronted building that is still intact near the corner of McKeown and Third was the office for the yard of the Tucson-based Corbett Lumber Company.

The Corbett Lumber Company building is still standing. Unoccupied, it serves primarily as a billboard on which the Home Plate Restaurant advertises. Photograph (2005) by Robert Whitcomb.

There was a "Power Plant" just north of the spur, operated by Jack Colquette. It sat about where the Home Plate Restaurant is today, powered by an ancient Buick engine! I was never in town after dark, but I was told that at 9:55 every night the operator cut off the power for a moment, warning the town that the entire system would be shut down in 5 minutes. I don't know what use most of the people in town would have had for power when it wasn't dark! You may well ask: "What about refrigerators, 'swamp coolers,' or electric clocks?" None of these had come into common usage in 1928 or '29. The universal coolant back then was ice.[3] "But how could they run the bars?" What bars? Arizonans endured Prohibition from 1915 to 1933, so there were no bars open. Bootleggers, however, operated handily under the flickering flames of kerosene lamps and lanterns.

By the time I returned to Patagonia in 1946, the town's needs for power were being served by Citizens' Utilities, operating out of Nogales. The Citizens' stub line—from Flux Canyon—is in use even today as far west as Costello Drive and

---

3. At the commercial level, "aqua ammonia" was used as an ice-making coolant. See Volume 1 for more information on this type of cooling.

still furnishes power for the area where the "Turkey Shoot" Rifle Range now sits, in the mesquite grove east of the cemetery.[4] Sulphur Springs Valley Electric Cooperative serves the area to the north. In downtown Patagonia, even as late as 1948, there was just one street light, and *it* was privately funded by Dawson Scoggin, who put it in to facilitate his business, the Big Steer bar.[5]

Richardson Park, on Fourth Avenue between McKeown and Duquesne Avenues, immediately southwest of the Community Church, was given to the Patagonia Volunteer Fire Department circa 1917 by R. R. Richardson. In that same year the cemetery, which lies within the San José de Sonoita Land Grant,[6] was given to the Patagonia School Board by the grant's owners. Inasmuch as there were no other governing bodies in the town at that time, these were the only available organizations to turn anything over to. Both tracts are still in daily use by the general public; the cemetery was turned over to the town in 1964.

For many years, important public gatherings were held in the Opera House, which stood on McKeown Avenue between Third and Fourth Avenues. It was built circa 1900 and was torn down circa 1950. Its main room boasted a dance floor, which was well used in its early days. There was an elevated stage used for plays and other productions, and there were provisions for showing movies. In its later years it served as a high school basketball court.

---

4. This grove was the site of a U.S. Army World War I campsite used by troops who guarded the border as an overnight stop on the march between Camp Nogales and Fort Huachuca. I believe that the several machine-gun pits that were dug on the edge of the cemetery hill to protect this encampment still survive.
5. The Big Steer went through various ups and downs and finally closed for good after the death of its then owner. It has since been completely renovated and converted into La Misión de San Miguel, a smoke-free night club and cultural center that features a variety of local and national musical groups.
6. In the early and mid-1800s Spanish officials gave a number of large land grants to favored Spaniards. The San José de Sonoita Grant was situated on both sides of Sonoita Creek and was designed to include the choicest riparian lands. Following the Gadsden Purchase, the U.S. Land Court upheld the legality of 7,592 acres of this grant (Barnes, 1983). Land grant issues in the Southwest were resolved on a case-by-case basis, resulting in some major conflicts (Montoya, 2002).

Patagonia Opera House, circa 1950. Photographer unknown. (Courtesy of Bill Bergier.)

Membership in the Volunteer Fire Department gave men high status in the community; in fact, this is still true today. The VFD's New Year's Ball at the Opera House was the social highlight of the community for many a winter. The firemen's principal fund raiser was the Fourth of July Rodeo and Barbecue, held at the Circle Z Ranch. For this event, they butchered six to eight beeves on at least some occasions and usually served some 1,000 attendees. I attended both the 1928 and 1929 barbecues and am proud to have been a Volunteer Fireman for six to eight years, beginning circa 1946.

In 1928-29 Patagonia provided church services for members of the Catholic and Protestant faiths, as it does today. The Patagonia Community Church was finished in 1928.[7] At that time, the Catholic community was served by a tiny church (20' by 40') that had been built in 1901 as a mission of the Nogales church. In 1931 that church was replaced by St. Wilhelmina's, but it was still a mission church served by a priest who came up from Nogales once or twice a week.[8] There were

7. Construction of the church began in 1917. The building stood unfinished for a long time. Patagonians are patient but persistent!
8. In 1955 the current Patagonia Church, St. Theresa's, was established as a parish in its own right. It now has its own mission church—Our Lady of the Angels in Sonoita.

also a good many business establishments. The Patagonia Commercial Company, a big adobe general store situated where the Stage Stop Inn now stands, was run by Val Valenzuela (1851-1945). There were several other groceries in town, one of which was operated by pioneer A. S. Henderson (1859-1944).

At least two groceries were run by Chinese men. These stores followed the south-of-the-border custom of awarding *pilón*, whereby children, assigned to trot across town to make small purchases for their parents, would be given candy or other small favors by the shopkeeper as tokens of appreciation for their faithful patronage. There were still two or three Chinese males in Patagonia as late as 1947 or '48, one operating the Henderson store under lease from Eva Stevens Henderson, who survived her husband by more than 20 years. Interestingly, in addition to owning the grocery, Mrs. Henderson was a practicing chiropractor.

The Washington Trading Company (named for Washington Camp, near Lochiel) was another business on the south side of McKeown Avenue. It carried a large stock of groceries and other goods and was a strong rival of the Patagonia Commercial Company. Both stores were built a foot or more above street level, no doubt as a precaution to protect them in times when floodwaters roared down Third Avenue from School Canyon.[9] Such flooding also led to the construction of a series of arcades and sidewalks that still survive along McKeown west of Third. Joker Mendoza worked on these as a WPA employee (Mendoza, 2001). There was a confectionery where the gas station now stands on the southwest corner of Naugle and Third. It was operated by a widow who had a soda fountain and also sold magazines and books. I recall discovering the *Readers' Digest* for the first time in the racks there in 1928. It cost a quarter, which was a significant amount of money in those days.

The Commercial Hotel at Naugle and Third, where the Patagonia Market now stands, was a big adobe building that had a dining room as well as rooms for rent. A separate "annex" was connected to the main building by an elevated, roofed wooden walkway. The annex operated until 1943 or so and then became Dr. Delmar Mock's first office in 1946. However, it was torn down, adobe by adobe, circa 1950. The adobe bricks were then used to construct an adjoining residence.

---

9. Flooding from this canyon, earlier called Barnett Canyon, was mitigated by the CCC in the thirties by counter-erosion work.

Ray ("Buck") Blabon and his brother Bert operated the East Side Garage. Buck, who came to town with his wife, Hilda, in 1915, had been master mechanic at the 3R mine, which was named for town founder R. R. Richardson.[10] The Blabon business was split into two sections, one on each side of McKeown Avenue, immediately north and east of the Community Church. A garage in which Bert served as principal mechanic sat on SP ground on the north side of the street and had a gas pump. The service station, on the south side, also sold gasoline and oft-requested automotive supplies. Among these supplies were Standard Oil petroleum products, furnished by a "bulk plant" set up alongside the railroad tracks just opposite Richardson Park. The bulk plant was operated by Rue Capps. His stock in trade was delivered to his door on a railroad spur.[11]

---

10. R. R. Richardson and Scott McKeown were reported to have been financially associated with John D. Rockefeller. When Rockefeller founded Standard Oil at Franklin, Pennsylvania, the story goes, he did so with the financial assistance of two younger men, Richardson and McKeown. Richardson retained an office in Franklin. (There are numerous names on the Patagonia map that were derived from Pennsylvania names.) Those two later sold out to John D, as he was known throughout the country, and headed west with large amounts of cash to increase their fortunes. Both acquired local land—McKeown by "homesteading" and Richardson by purchase of a considerable number of tracts. We know a great deal about what became of one of Richardson's many enterprises—we call it Patagonia. But the details of what McKeown was up to locally or even what became of him are obscure. Some think that he "trod the primrose path," a term first used by Shakespeare to refer to a pleasant path to self-destruction.
11. Although in the heyday of this SP line there were two or three trains each way every day, service was eventually provided less often. During World War II the train operated only three days a week. The trains were "mongrels" that carried both passengers and freight. A typical train of those days might have had a locomotive, a passenger/mail car, an oil tanker, several ore cars, and several cattle cars. In the passenger/mail car there were seats for about 30 people.

East Side Garage, operated by Buck Blabon from the 1920s until the 1960s. Photographer unknown. (Courtesy of Bill Bergier.)

Standard Oil's gasoline (Red Crown) could be bought not only with cash or credit but also with coupons, issued in "tear-out" booklets that were worth various fractions of a dollar. This form of currency was locally called script, although its true name was scrip, as proclaimed by a handsome enameled metal sign alongside the pumps that advertised "Scrip Accepted." When we came to town during the geology field course in 1928, we had an ample stock of tear-out booklets to keep our U of A Chevy pickup in operation.

There was a weekly newspaper in those days—the *Santa Cruz Patagonian*; it was published from 1912 to about 1930. The newspaper office was on Third Avenue, in what is now the annex to the Seventh Day Adventist Church. Howard Keener was the owner-editor of the paper. He also had a 160-acre homestead in the San Rafael Valley, west of today's Vaca Ranch.

More or less in the middle of town, the SP maintained a complex of sturdy wooden shipping pens, complete with facilities for feeding and watering stock that was being held for shipment to market. The facility had wooden ramps up which the animals could be driven into the cattle cars. The pens were just opposite the Community Church, on the north side of the tracks. Joker Mendoza, who lived across from the pens, remembered that some Brahma bulls once got loose, tore down his fence, and ran around in his yard (Mendoza, 2001).

Outgoing ore shipments were dumped directly into gondolas opposite these structures from trucks that backed up steep ramps (made of earth and heavy timber) tight against the railroad sidings. LCL (less than car load) shipments of concentrates or ores of higher "tenor" could be off-loaded (shoveled by hand) from wagons or trucks[12] onto massive plank platforms built up high enough to match the level of the floors of boxcars. After the gondolas or boxcars had been "spotted" opposite the piles of these materials awaiting shipment, they could be loaded by hand pitching the ore—hard work!—into "orbits" of impressive heights, a shovelful at a time, or wheelbarrowed into boxcars from the plank deck of the long wooden ore dock.

Patagonia Commercial Company in the 1920s. Note display rack for local ores at right. From a postcard of that ear. Photographer unknow. (Courtesy of Bill Bergier.)

In 1928, the standing of Patagonia as a mining center was advertised to any casual passers by who might somehow be unaware of that status. There was a tall 4" by 4" post that looked a bit like a hat rack set into the ground where Duquesne Avenue meets Harshaw Road. It bore horizontal 1" by 4" signs that displayed the neatly lettered names of 12 or 15 nearby mines, giving distances as well as pointing out directions. Oddly, I have been unable to find anyone else who remembers that sign. I don't *think* I just imagined it! Several similar signs exist today, farther out Harshaw Road, but they point to ranches. There are no longer any active mines to point to.

12. Freight wagons, still common in those times, were also used. Many were owned by the Ahumada family. Some members of this family still live in Nogales.

Every now and then an ancient postcard turns up that shows the front of the Patagonia Commercial Company and some of McKeown Avenue in front of it. In the picture, you can see a set of racks or shelves made of whitewashed lumber, with 1" by 4" sides surrounding the top as well as all the shelves. Not readily apparent in these photos are large "samples" of ores being displayed in the racks, along with the names of the mines that furnished the specimens. Of course, the samples were not high-grade ore. They didn't have to be high grade to demonstrate the types of materials being extracted from specific prospects or mines. A lot of these samples were carried away, but new ones were always brought in to replace them. The display made for good public relations for the town and bore no real risk. If somebody carried away a 5-pound chunk of low-grade gold ore, they might have had 70 cents worth of gold—if they could crush it, process it, assay it, and get a smelter to accept a 5-pound shipment. Pretty cheap public relations!

A display stand showing ores from local mines (the names Duquesne, Wandering Jew, and Hardshell are legible). This stand may have succeeded the racks in front of the Patagonia Commercial Company. Photographer unknown. (Courtesy of Bill Bergier.)

Alongside the SP tracks and these display racks was a huge piece of timber—about 8" by 8" standing 12 to 15 feet in the air. Atop this impressive post was a contraption that looked much like an old-fashioned coffee mill. But a mill it was not. It was the town's fire siren, with its on-off switch mounted on the pole, 6 or

7 feet above ground level, high enough so that small or middle-sized children couldn't reach it. I never heard of this happening, but surely at some time mischievous kids must have carried a stool or something to stand on and set the siren off at night. And just as certainly, I am sure they caught hell for it! When the siren was turned on, it gave a loud "roaring bull" sound. It was such a fine and wide-reaching sound that it was also used to announce the 9 p.m. curfew, which was meant to warn children of school and preschool age to go home and not get caught running around unescorted.

C. A. Pierce, a graduate of the Missouri School of Mines, owned an assay office on the southeast corner of Third and Duquesne, where he analyzed samples brought in by miners and prospectors. He had been mine superintendent at the Mansfield Camp and was the longtime owner of the "Victor" lode lying just east of the Baca Float, west of town.[13] He made reports on mines, as well as surveys and maps of land, town lots, and surface or underground mining properties throughout the area.[14]

The original Patagonia Grammar School was a small adobe building just west of Second Avenue and north of Pennsylvania Avenue. It was later converted into a residence. An impressive brick mission-style school was built on the mesa in 1914 and was one of the town's prominent buildings when I was there in 1928 and 1929. It has had various additions since then. This structure still serves its purpose well.

---

13. A float is a government grant of a fixed amount of land not yet cut up by survey out of a larger specific tract. Luis María Cabeza de Baca was a descendant of Alvar Nuñez Cabeza de Vaca, to whom the King of Spain had granted a large tract of land in what was then New Mexico. Later, land in Colorado and Arizona was added. Further grants were made by the Mexican Government in 1821 to heirs of Luis María (Gorby, 2000).

14. Pierce also served as State Senator for Santa Cruz County when I first knew him. His daughter Sarah (Sally) was a year or two behind me at the U of A and eventually became a well-known archeologist.

Patagonia Elementary School, built in 1914, occupies a site once used by Indian lookouts. Photograph (2004) by Robert Whitcomb.

There was but a single high school in all of Santa Cruz County, in Nogales, until about 1928, when an old frame Empire Smelting Company warehouse on the southwest corner of Third and Duquesne was converted into the Patagonia Union High School. This building, along with several auxiliary buildings, served the east end of the county until circa 1947. The entire complex has since been dismantled. Its first principal was a young man, Ralph "Red" Zimmerman, whose tenure is commemorated by the sole remaining piece of this high school structure—a 2-foot-high retaining wall of rubble masonry built, by way of restitution, by miscreant male students over the course of several terms. It was originally 130-odd feet long but is down to 50-60 feet at this writing. It is reported that the only expense involved was the cost of the cement used to mix the supporting mortar. All but the largest stream-bed stones were carried by hand to the work site by miscreant female students!

Detail of a rock wall built by miscreant students during the time that "Red" Zimmerman was Principal of the Patagonia High School in the 1930s. Photograph (2005) by Robert Whitcomb.

In 1947 a new high school was imported and placed east of First Avenue and along Route 82 on the grounds of a former U.S. Forest Service headquarters. The building wasn't fancy. For $3,500, the School Board bought a large two-story frame building from Fort Huachuca, where, in its heyday, it had served as a recreation center. I watched with fascination as the rec center morphed into a school. The building was T-shaped and had overall dimensions of 200 feet (the top of the "T") and 100 feet (the vertical leg of the "T"). Before they could transport such a large building, the movers had to saw it into seven sections. The buildings were trucked 40-odd miles to Patagonia, where they were set down atop matching foundation pylons. The move cost about as much as the building, and the pylons and patching of the saw cuts cost another $3,500. The total cost of the building was therefore less than $12,000. The makeshift structure, after reassembly, had good structural integrity and served its purpose well until being replaced by a larger structure. As of 2005, the high school is undergoing expansion again. In the old days, it was the rule, rather than the exception, to "just make do"—to find an inexpensive means to accomplish a necessary goal. Today, people just assume that anything worthwhile costs hundreds of thousands or even millions of dollars!

Today as I look at Patagonia, I sometimes remember the way it was back in 1928. Of course, that was 77 years ago! The town has changed, but I think it has gotten better. Essential services have improved. The volunteer firemen of today have modern equipment and have modern response units for emergencies. There is even a helipad so that local people needing emergency health care can be rapidly transported to Tucson. Actually, I took advantage of this service myself a few years ago when I fell down my concrete stairs holding an armload of maps. I recovered fast, given the good medical help, and, just as important, the maps all survived too!

However, in ways that affect everyday life, the town hasn't changed much. There is a corner market where we shop for groceries, and a single service station, just as there was then. As I drive down the streets I see a few very, very old adobe houses—dating from the 19th century. As I mentioned, some houses—like the old railroad section house—were moved intact from one location to another. Others have survived but have been renovated so much that they no longer are recognizable as I remember them. For example, for a number of years the Mesquite Grove Gallery has occupied the home that was R. R. Richardson's residence in the early years of the town.

An old adobe home on Sonoita Avenue, probably built in the late 19th century. Photograph (2005) by Robert Whitcomb.

There are perhaps more changes in the countryside around Patagonia than in town. New homes are springing up on the (Red Mountain) Mesa, and on Gringo and Salero Roads. In the narrows of Sonoita Creek, the sheer face of rock that was quarried from time to time over the years occasionally sheds some loose pieces. The Department of Transportation recently provided remediation from these periodic rock falls. Years ago, a rock quarried there was used in drilling competitions. That rock now stands in front of the old depot, which survived a road widening project in 1964 by being moved 50 feet. The quarried contest rock now bears a plaque designating the depot as a historic site.

A rock, quarried at the site of the Highway Rest Area, that was once used in drilling competitions today bears a plaque designating the Patagonia Depot as a historic site. Photograph (2005) by Robert Whitcomb.

In all, I think Patagonia has escaped modernity pretty well, and for me, it has provided exactly what I was looking for when I decided, upon coming back from World War II, that I was going to make it my home.

Coming Back

In 1945, at long last, the war was over, and I was duly appreciative of that. I had left for the service at a rather pivotal point in my life. I knew I didn't want to have a boss other than myself. But I wasn't sure what I should be doing. The scheelite mill had gotten me through hard times, but it hardly promised a career! I was unsure enough of my career path that I actually went to an aptitude testing center in New York City, where I was run through a whole battery of tests. For all I knew, I should have been a pilot or an English teacher or a politician. I really wasn't sure. But the tests said no, you belong in engineering.

Several of my old friends in Yuma urged me to come back there. They said, "There's plenty of good salaried jobs for you here, where you know your way around." But I told them, "It's too hot!" They said, "We have air conditioning." I replied, "Are the streets and lots air conditioned these days? You know surveyors work outside." They said they hadn't air conditioned the out of doors, so that ended that! I decided to return to Patagonia after a 17-year absence.

The Mines Weren't Operating

Unsurprisingly, after five years in the Army it took me a while to really feel like a civilian. Before I settled down in Patagonia in March 1946, I picked up my things that had been stored in Yuma. Among them were five pairs of good Levi's, making me the best-dressed gent in Patagonia for quite a spell.[15] Unable at first to find a rooming house, I rented the rear bedroom of a house from Mrs. Neil McDonald, the widow of an old-time miner.[16] The room had a back-door entrance, so it was convenient. This house, across Naugle Avenue from the present-day Post Office, is now Grayce Arnold's candle and gift shop, a local landmark.[17]

---

15. Patagonia is noted for informal attire. Some say the official garb is "Patagonia grunge." I never really donned the "official uniform," but, like all Patagonians, I have enjoyed my "Levi's uniform" in the town.
16. Mrs. McDonald's son, John, was a longtime manager of a large graphite mine in Sonora and was very well thought of in Patagonia.
17. Grayce Arnold is the oldest shopkeeper (in both age and business longevity) in Patagonia; in March 2003 she celebrated 25 years in business.

After only three or four nights with the McDonalds, I was able to rent a room on Pennsylvania Avenue from Charles Mapes and his wife. I mentioned him earlier. The Mapeses' house was an old adobe. They converted the front porch into a room for me by enclosing it with a shellacked window screen, which sealed it in while admitting a lot of light—plenty for drafting. I converted that makeshift room into an office, where I kept a drafting table and other mapping and surveying equipment.

I had come back to Patagonia high on hopes and expectations. I had assumed that I was returning to a really active mining community, and I planned to take on a lot of mining business. After all, when I had been in Patagonia in 1928 and 1929, there had been 20 or 30 little mines within 10 or 12 miles of town. I was sure that some of those mines would have remained active during the war because of the demand for copper, lead, and zinc. I didn't even bother to check it out. There were 15 or 20 people around the area that I had known and could have checked it out with when I was overseas, but I didn't.

My intention was to establish an office with engineering and assay capabilities. In addition, I was prepared to expand my services to other mining-related activities, if necessary. I also planned to reinstate my commission as a U.S. Mineral Surveyor[18] and to take on surveys of mining claims for patent.[19] In addition, I planned to do mine surveying, mapping, and reports; to stock powder, caps, fuse, carbide, timber, and other mining supplies for sale; and to take small shaft-sinking and tunneling contracts. I expected that, with the high wartime prices of metals that I knew had been prevailing, every little mine would provide business now and then. And some little mines, I figured, were bound to have become larger. So I was ready to do wonderful things.

Unfortunately, I had a big surprise waiting for me. The mines in the Patagonia area had become marginal, and most of them had been shut down. As the fortunes of the mines declined, their managers were unable to pay miners wages even remotely comparable to those paid by large, successful mines. The few miners left in the Patagonia area were getting only $6 or $8 a day, while the big mines like those in Morenci or Globe were paying $10 or $12 a day. The wages weren't quite that high in Bisbee, but all the big mines in southern Arizona were working to capacity. So mine closures were removing the single industry that had supported growth in Patagonia. As soon as potential miners were old enough to

18. I had taken the six-day exam for this commission in Phoenix in 1933.
19. Patented mining claims are those that are in compliance with laws governing claims and that have been legally recorded and convey full ownership to the patentor.

work—if they hadn't been drafted—they could go to work in the bigger mines and enjoy more perks while also making a lot more money than they could in Patagonia, if they could find a job there at all.

I Knew I Was Home

So my plans had "gone agley" in that first year. I wasn't sure I could make a living in Patagonia, and the winter days were beginning to weigh on me. Then one early spring evening, when I was feeling kind of low, I opened the window and heard some Hispanic neighbor kids playing a game and singing:

Go een and out the guindows,
Een and out the guindows,
Een and out the guindows,
As we have done before.

They were singing in English, but there is no "w" in Spanish, so "window" comes out "guindow." I had heard that song often in Yuma during my childhood, and here it was in Patagonia. I knew I had come home. I would find a way.

Mining Again

Setting Up a Business

After I got out of the service, I took a train to Omaha, where my parents had moved for Dad's medical care after he found out he had leukemia. I spent a month with my parents and then retrieved my car from them (they had an order in for a new car to replace my 1938 Chrysler) and drove back to Bisbee for a bit. From there I ranged around with some of my old prospecting pals. Fred Montgomery had lived near Alto for some time, so some of our ranging around was out that way. Local Patagonians will recognize Alto as the site of the late Ray Bergier's ranch, and mining buffs will remember it as the mining country in which the Salero legend arose (see page 138).

When I set out to begin my business I had two potential partners in mind—Greg Brickhouse, who had been the motor officer in the 1318th Engineers, and Jack Sylvester, a civil engineer with whom I had worked on the All-American Canal. However, as it turned out, neither of them actually joined up with me, so I had to go it alone. I bought good surveying equipment and got several underground mine survey jobs right away. For example, I picked up the survey of the nine American Boy[20] lode claims for patent. I also surveyed a few ranches. But these jobs—all of them put together—weren't enough to keep me going. I felt that I would have to become a miner myself, if nobody else would. Hugo Miller's assay office in Nogales was already taking very good care of what assay business there was, and it was clear that there wasn't much mining work left for me to do. I'd get a little job on occasion, but I also took time between jobs to go and visit old friends and try to build up a little confidence. I looked at quite a few small mines and prospects trying to find a deposit that could be worked profitably. Like everybody else, however, I had a hard time finding anything that promised to be economically viable.

Scratching a Living Out of the Rock

Most locals felt that the pay dirt in the Patagonia hills was all mined out. However, for lack of any other business, I had no choice but to look for some. I was going to have to scratch my living out of the earth. Over the years I had done quite a bit of primitive mining—hand drilling, hand loading, and hoisting with a windlass. I never was very good at hand drilling, but I did know how to drill and blast, how mine timbering should be done, and so forth. Actually, I was more involved in hauling, buying, sampling, and shipping ore than in literally digging it out.

It occurred to me that there was a circumstance that I could exploit. Most miners back then had worked in several mines—maybe a lot of mines. It was typical for them to remember that, years earlier, there had been a little streak of ore—maybe only 3 or 4 inches wide—left untouched in this shaft or that tunnel. The mining companies no doubt knew about most of these streaks but bypassed them because there wasn't enough value there to make it worth their while to mine them. But the hired hands would later say to themselves, "Well, back in 19 so-and-so and in the such-and-such mine, we left a few crumbs behind. Let's see if we can reopen those workings and sample them; those leavings just might lead us to ore!"

So they would go into one of the old mines. If there was enough ore left to be mined manually, they would do that. In a week or two a pair of hand miners could gouge out a ton or two of ore and make a profit if its value (in silver, copper, and/or lead) was high enough—say $50 to $100 a ton. If a big company blasted it,

---

20. The American Boy Mine group is in the Wrightson District, Santa Rita Mountains, Santa Cruz County, Arizona. It was worked as a gold/silver mine with copper values from about 1906 through 1918 by partners who lived on the key claim and sporadically thereafter through 1951 (Keith, 1975).

everything would be shot into the "muck pile" and would be scattered through a hundred feet of tunnel. The values would be diluted by the waste rock, and the quality of the ore wouldn't be nearly as high as if it were carefully hand picked out of the streak. If the little guy blasted it at all, he could do it in a very targeted way, so he wasn't tearing up the whole countryside to get some ore. What's more, he didn't have the resources to blast, even if he wanted to. He could hardly afford the dynamite, let alone all the equipment it took to run even a small operation!

The Old Hand Drillers Are Gone

There used to be quite a few miners in town who did a lot more hand drilling and hand work than I ever did, but they are almost all gone now. With the passage of Joker Mendoza (1917-2003) and Fred Benedict (1909-2003) it seems appropriate to remember, in these days of mechanization, how hard the early miners worked. They really knew how to do things the hard way—it was the only way! During the Depression, my friend Hugo Miller was involved in cleaning up stray pockets of ore in old mine workings. Of course, operations of any size, even back then, had compressors and other equipment. But in a long tunnel, if you found a little batch of ore—a little stringer that the big miners had not worked back in the boom days of World War I—it might not be worthwhile to put in pipelines for water and compressed air. You could do cleanup work by hand without these "improvements."

Fred Benedict once told me about a time when he was working underground at the 3R Mine, which Hugo was leasing. The work was in the hardest kind of granite. The miners were drilling to retrieve copper ore that had very little silver in it. At the end of each working day, Hugo would come around with a yardstick and measure the holes that each man had drilled. After a long shift, a miner might have drilled 3 to 5 linear feet into the granite. For that he got 3 cents an inch! That was not a lot of money—even then. Hugo would figure up the total and pay off out of his pocket, right then, in cash. If a miner didn't show up the next day, that was OK. At first, Hugo blasted the holes himself, but later Fred helped him, and he drew extra money for that. Fred didn't tell me how often he took a day off from that exhausting work, but since it was a time when he needed money—like everyone else—I imagine that he didn't take many days off.

Paul Conley

Such circumstances created an opportunity for me, and as a result I often had a little mining "deal" going on the side. I was what they call today a "small businessman." Real small! Small businessmen breaking into an untried resource used to be called entrepreneurs. It was natural for me. I learned thrift from my parents, and I always saved whenever and whatever I could. Like all small—tiny would be a better word—businesses, I cut corners when I could. With different partners, I located and worked marginal deposits of copper and lead and found opportunities to buy ore as a result of association with other tiny businessmen.

To get started, I "anted up" for a one-ton Model B Ford manual dump truck for $100. Later I expanded and improved our operation by buying a 2½-ton GMC (power) dump truck for $250. On weekends Paul Conley, a full-time ore-haul dump truck driver for a local firm, Strong and Harris, drove one or the other of these trucks for me, hauling ore to railroad cars to be shipped out.

Paul was a good friend and a hard worker, but he had no mining experience. I had very little hauling experience, so we made a good team in some respects. I once drove the smaller truck backward up one of the big railroad loading ramps on Naugle Avenue, but I never even tried with the GMC. But even with the two of us, it was sometimes difficult to follow projects all the way through. Because Paul had only weekends to help me, he was limited in what he could do. We often worked together sampling bulk ore to buy from small miners, a process that required a lot of shoveling.

Paul and I established a different line of business that was more suited to our respective experiences. We set up some copper-leaching troughs, one each in 3R Canyon and Mansfield Canyon above the American Boy, and reclaimed copper from surface water. We would put bins on the ground outdoors, where copper water was flowing; fill them with all the scrap iron we could find; and run the water through them. The fancy term for these troughs was "precipitation tank." I described this procedure as it was done in Phelps Dodge mines on page 21, in the Bisbee chapter. Paul and I got maybe a pound or two of copper a day per trough in each of our tanks. Using a square-point shovel, we would shovel the contents of a tank into a wooden powder box and just let it drain. We couldn't get all the iron scraps out, so we would put the box—scraps and all—in with the next copper ore shipment we made by rail to the smelter. Our carloads would be thoroughly mixed at the smelter before sampling, so we would recover our values in the form of a bit higher copper content of a carload of ore than it would otherwise have had. Paul and I worked together on the precipitation tanks for six years. He and friends continued to work the tanks even after the Conleys moved to Tucson.

Working With Lalo

In about 1948, I began working with Lalo Vasquez and Leland Wilson. Lalo was a good worker, knew the mining business, and was easy to get along with. We became good friends. I anted up the cash, and he kept the work going. In what I believe was our first job, we shipped two carloads of copper ore from a lease we had on a dump southwest of Harshaw. Lalo picked up three or four pack burros off the street in town and made pairs of pack boxes (I still have a couple of pairs). He then packed the ore out of the steep country to a temporary dump site. I then paid Lee Kuhn for a quarter mile of bulldozer work for a new side hill road from Harshaw Road to our temporary "stockpile." When we had it all gathered up, we trucked it to town and shipped it to the smelter.

Lalo and I worked together for a long time. In about 1960, we scraped together two railroad car loads of silver "leavings" at an old Mowry smelter site, plus eight or 10 tons of similar ore at an old mill site in Harshaw. Because the leavings were on private lands, we had to pay royalties to the owners—a percentage of the returns. Of course, we had to wait to pay the owner until the railroad tonnages had been determined. In this case, as was customary for me, I paid only 10 percent of the estimated value. Even so, we were working on a very narrow margin, so we had to sample carefully to estimate the values. Of course, it had to be sampled and assayed again by the smelting company's assayer.

An unusual and interesting circumstance arose with the Harshaw shipment. The site from which we recovered the silver leavings was a disputed claim. There were overlapping boundaries. When we were in the middle of the job, we found out that Henry Acevedo might have a prior claim to the land the leavings were sitting on. So I worked out a deal with Henry. I surveyed his claims in the area as compensation—if there was a need for any—for silver leavings that might have turned out to be on his land. In the long run, the survey showed that the leavings had never been on his land. But we avoided a confrontation and were both satisfied.

More often than not in those days, we worked without contracts. Generally, there were three smelters that we used. Ore with primarily lead value went to El Paso. Copper ore went to El Paso, Douglas, or Inspiration (near Globe). On one occasion, which I remember as being especially informally conducted, we didn't ship to any of the usual smelters. Lalo and I shipped a carload of lead ore to the Selby Smelter, north of San Francisco. The ore was from the Dixie, an unpatented lode claim a short distance northwest of the American Boy. I knew that there was a nice body of galena ore there, off a short adit. The ore was in a shallow winze that was full of water. This called for outside help, so we made a deal with Strong and Harris to do the haulage. We rented an air compressor and drills, plus a "sinker" pump. Strong and Harris reworked a short length of road and delivered all of our equipment to the mine site. Once we had the pump there, it was a snap. We dewatered that winze while we ate our lunch! Then we mined the ore and Strong and Harris hauled it to town. It took several weeks before we were paid off from the Selby "returns," so everyone involved had to wait for their money. But nobody was unhappy about it. All of those transactions were performed on verbal "contracts." My goodness! You couldn't do that today—anywhere.

The Shipment I Didn't Make Money On

Now, as I mentioned earlier, my various mining partners and I didn't ordinarily do the blasting and mucking—but in several instances we did get a lease on a property and mined the ore ourselves. I remember one occasion in about 1960 when we put together two carloads of copper ore that way. They were from the fringes of the 3R Mine, which is in 3R Canyon, across the highway from the Circle Z dude ranch. We hired a dozer and graded a new steep switchback road on a hillside. That was one of the few times we didn't have our own truck, so we had the ore hauled to town by Strong and Harris trucks. I was depending on getting 5 to 7 dollars a ton in silver, but those carloads ran almost no silver or gold. I was lucky to break even on them—actually, as I recall, I was about $100 ahead. Needless to say, we didn't mine any more from that deposit!

Shipments I Did Make Money On

The 3R bust was the exception. I often managed to make a little bit of profit in my other ore-buying transactions. I would often be working on a carload of copper ore and one of lead ore at the same time. Ordinarily, the copper ore in the Patagonia region would "carry" 5 to 10 dollars worth of silver and gold per ton. That was only $150 to $300 total, of course, but that would generally pay the freight and would sometimes pay the smelter's treatment charge for the values. Luckily, I got into this business during a period when the price of metals was on the upswing, so there was a little bit of profit. Once in a while I'd get a little short on capital because it would take 30 to 40 days from a shipment date to get my money back from the smelter.

On occasion, I ran across some ore of sufficient value that it was prudent for me to safeguard it. I felt that it might not reach the smelter if it were left unguarded for several dark nights. Before my time, when someone had a lot of such valuable ore, entire 30- to 50-ton carloads would be loaded in a single long operation. Such an operation called for the use of hand-loaded wheelbarrows "trammed" across a 2' by 12' timber plank or two that served as the dry-land equivalent of a gangplank. In my case, I always kept the highest grade ore inside my office on McKeown

Avenue. I never had a whole carload of it—that's for sure! The normal "mine run" ore was allowed to pile up in separate lots, on the railroad's big downtown timber ore dock, as I bought it. Trains normally came in from Benson via Fairbank on Mondays and Thursdays. The gondolas that were on Thursday's train would just sit on a siding until the weekend, so we had until about noon Monday to load them, without demurrage.[21] On this branch line with its light steel rails (standard-gauge), the railroad company would accept 30-ton carloads without penalty. On the main line, they always had a 50-ton minimum charge, which we would have had to pay even if our shipment was less than 50 tons. It therefore behooved us to arrange for at least 30 tons in all our shipments.

Although ore buying wasn't a big money maker, it helped me make a lot of good contacts. I bought high-grade ores and over time was involved in the shipment of something like 25 railroad carloads of direct-smelting ores.

One factor that prevented any large-scale expansion of my mining business was a legal restriction. Just about all U.S. citizens, or even people who merely declare their intention of becoming citizens, can stake claims under the mining law of 1872, but I couldn't. Because I was a U.S. Mineral Surveyor, I was kind of a "pseudo employee" of the Department of the Interior and therefore couldn't stake claims. So I couldn't have become a "mining magnate," even if I had wanted to. What I could do, however, was lease unpatented claims or buy claims already taken to patent by others.

The Amazing Downhill Transport Invention

There were a lot of us trying to scratch a living out of the rock at that time, and we were always looking for some way to make it just a little easier. I remember one innovation that seemed to be a very good idea at the time. An acquaintance of mine here in Patagonia was mining up in the Salero area. I think he might have found ore on somebody else's claim and needed to get it out fast before the owner found out what was going on. I never asked him for the exact details. In any event, he had sorted out a lot of ore uphill from the mine road. He didn't want to build a road up to the loose ore, but he needed to get the ore down—fast! He finally concocted a plan. He packed some 55-gallon drums with lids uphill,

21. Railroad conductors set out cars on sidings or spurs "on order" for a stated period deemed sufficient for loading. If the car is not loaded when the time is up, a demurrage charge is added by the hour or the day. In Patagonia, with only two trains a week, we had three or four days to load up, so we did not have to worry much about charges.

where the ore was. Then he filled them with ore, fastened the lids on, laid them down, and rolled them down the hill. But the drums went faster and faster as they rolled, and about a third of the way down they broke open and scattered ore all over the hillside. That place probably is still upholstered with low-grade ore! He couldn't reuse the drums, either, because they were bent and dented. It was a real good idea—but it didn't work!

Clio Walker

The best prospector I ever knew was Clio Walker. He was from New Mexico, but he lived in Patagonia in about 1946-48. I got to know him very well, and when he left, he sold me a 20-acre patented mining claim, the Iron Cap, up Alum Gulch, for $100. And he threw in a three-sixteenths interest in another claim, the Isabel, east of Tubac. I still have my interest in the Tubac claim and still pay my share of taxes on it. You see, I have never stopped mining. Not even now!

Clio was really into prospecting. One day he said, "Let's go out digging tonight. There's a full moon. We'll go and dig a hole. I know where there's some good stuff." "Now wait a minute, what's this middle-of-the-night business?" "Well, it's in the middle of the road!" So that's why one moonlit night two indefatigable prospectors went digging in the middle of a county road. There had been a mill maybe 50 or 100 feet uphill 70 or 80 years earlier, and when driving past I had noticed that nothing grew there. The reason was that mill tailing—which is often toxic to plants—had washed down from the mill. Clio showed me exactly where he figured we should dig—a place where the earth in the road was slightly yellow. We dug down a foot or so, took a sample to assay, and filled the hole back in again. Luckily, nobody came driving by. Our sample assayed 20 ounces of silver per ton, and at present prices that would be more than $100 a ton—good stuff. Now, finding ore in a strange place is one thing, but taking advantage of it is another. We never did figure out how to get the ore out of there and cash in on its value! Not only was it on a Mexican land grant—privately owned patented land—but it was also well within the right of way of the county road. So it would have been hard or, really, impossible to get a lease to mine there. To this day some of those values are still out there!

Clio was always telling stories. He knew where everything was, and he knew everybody. I would say, "Is there any place that you haven't been?" And he would reply, "Well, I have been pretty well over the Southwest and Mexico." One time I drove to Magdalena over in New Mexico—that's a good mining area—to look at some property with my old prospecting buddy Ben Harden, from Bisbee. We visited a ghost town there that Ben remembered. He said, "Right in here was a little camp; it's long gone, but they were mining zinc ore there for the Sherman Williams paint company 50 or 60 years ago. They called it Hardscrabble." So I thought I had found somewhere Clio hadn't been. So when I came back I asked him, "Do you know where Hardscrabble is?" He said, "I was *born* in Hardscrabble!"[22]

Another time Clio asked me casually, "What's the best prospect you ever saw?" I said, "A-Vea-Hah [Moon Mountain], on the Mojave reservation, north of Ehrenberg.[23] In the south end of the reservation there's a shaft about 30 feet deep following a vertical quartz vein about a foot wide, and gold is sticking out all over the vein." I had gone out there with a fellow who had a legitimate 20-year lease on it. At first I wanted to take it over and mine it out, but then I changed my mind. So Clio said, "Why weren't you interested in it?" I said, "Well, down at the bottom, the ore was 2 feet wide." "Did that make it look good to you?" he asked. I continued, "But then I got to looking, and I found that the vein where they had sampled it was 2 feet wide, but the rest of the vein was only about 12 inches wide. At the place where they had shot into it, the vein had faulted (in other words, fractured horizontally) and slipped so that, in a short segment and only there, it was double width. I would estimate the double-width vein was no more than a few inches deep." Well, even with a 12-inch vein it was high-quality ore and was still worthy of mining, but it was only half as good as it looked at first. And I said, "That really got me." Clio said, "Yes, I know! I sank that shaft!" It had been 10 years earlier. He had sunk the shaft, but he didn't get the lease from the Indians. It had been awarded to a man I knew who lived close to the reservation.

SURVEYING

As time went on, I got busier and busier. In addition to buying and shipping ore, I started a survey business. And it was surveying that became my mainstay. Before I knew it, my problem was not that I didn't have enough work. Instead, I had too much! In 1949 I set up my office on McKeown Avenue, in the building where several businesses, including the Gathering Grounds café, Mariposa Books & More, Creative Spirit Artists' Cooperative, Global Arts, Darrell's Designs, and the 100th Monkey, are today. Sooner or later, I did just about every kind of surveying. Although I started out doing all the work with just one hired helper, I eventually set up crews that could survey without me once I made the preliminary drawings and reviewed the site with them. I always took solar observations to find True North and started the job off right. After that, the crews were usually okay.

---

22. Hardscrabble is about 300 miles northeast of Patagonia.
23. A-Vea-Hah is about 250 miles northwest of Patagonia.

Of course, I was responsible for the final mapped product, so I oversaw each project carefully. I'd check all their calculations and do the final maps, drawings, and reports myself. With my own survey crews, I made almost 700 claim surveys for patent. Over the years, I worked in every Arizona county except Apache and Navajo. I also worked in Grant, Taos, and Hidalgo Counties in New Mexico. Although I tried to stick to mining work as much as possible, I ended up doing a lot of general surveying in Nogales. I also did a good deal of mapping and property-line work in the rural portions of Pima, Pinal, Cochise, and Santa Cruz Counties, particularly when metal prices were down or mining work was slack

And, of course, I mapped. I compiled and drew (to be sold) copyrighted maps of Yuma County, Santa Cruz County, and the City of Nogales. The Lenon Engineering Office had microfilm (35mm on reels) of all sorts of records, including all mining-claim patent surveys made in Arizona up to No. 3000 or so; USGLO township plats on film for the entire State of Arizona (all counties); and all the Santa Cruz County deeds from 1899 up to 1960 or so. I still have all these maps. I don't know why I decided to take early retirement, but I did. I retired in 1975—I guess I thought I was supposed to. But I regretted that decision many times.

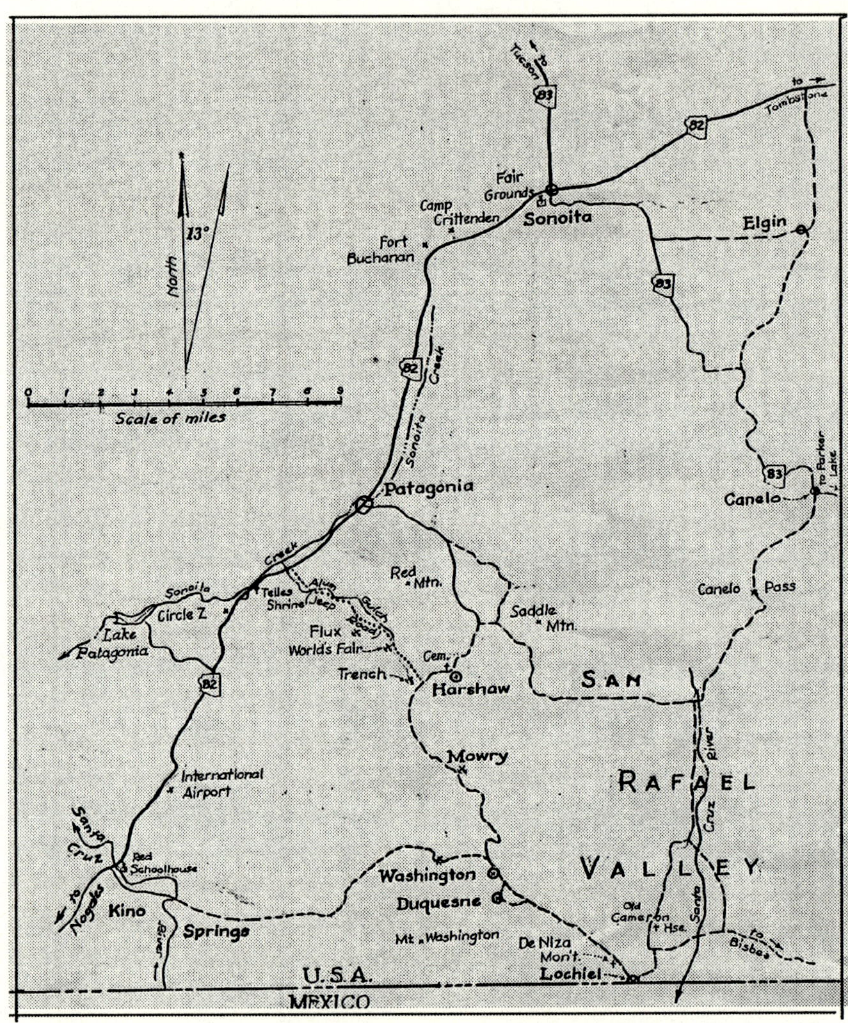

Map of the Patagonia region, excerpted from Lenon's map of Santa Cruz County.

My Valued Clients

Although surveying jobs were pretty routine, some clients managed to make themselves memorable by being royal nuisances. In going through my files recently, I ran across a list of local surnames on a business letterhead that I had made 40 some years ago. Entitled "Our Valued Clients," it lists people with whom I had dealings or, in some cases, whom I just knew casually but who had one thing in common—they were unpleasant to do business with. Some pestered me for follow-ups on jobs but never wanted to pay for the extra work. Some owed me money—and many of them still do, for that matter. A couple were guys who had worked for us briefly and didn't do too well. I didn't put in writing what the list actually was; various local women worked as my secretary, and I was worried that some of them might have been acquaintances of or even related to some of these "friends." So I didn't want to say, "These are guys that I don't want to deal with." I put their names down, like I say, as *favorite* people, "our valued clients." But if they called to pester us for something, the girl in the office knew enough to say I was out or something like that. I've gotten mellower as the years have gone by, but I haven't changed my mind about some of them! Of course, I have outlived almost all of them!

Here's an example of the kind of problems I had; this was in the late 1950s. One of my "honorees" was a neighbor of a friend and client. My friend had a little nursery business. The two neighbors wanted their common property line "run out" because they weren't getting along very well. So I staked out the line and mailed them bills for about $36 apiece. The fellow with the nursery paid up almost immediately, although I expect he had to kind of scratch to get that amount of cash. But the other guy didn't pay. After a month or two I asked for my money again, and he said, "Can you come by? Somebody knocked down one of the stakes." I had driven stakes on the line, but somebody had knocked one of them over. So I went by and reset the lost stake. I didn't really want to see the guy, but it did give me an opportunity to dun him again. This time he paid me on the spot, by check, signed: "So-and-so, Col. (Colonel) U.S. Army, Retired," which, even though I had been discharged as a first lieutenant, I endorsed: "Robert Lenon, Corporal, U.S. Army, honorably discharged," in order to be sure I was sufficiently humble. Then I deposited it. We had both served the Regular Army at the same time. Big deal!

That same "honoree" had 20 acres or so just north of Nogales, and eventually sold a piece of it to a movie star who was a longtime friend of his. So the two went out with a 50-foot tape and measured off what was supposed to be a tract 250 feet square. Well, they taped 250 feet fronting along a highway curve and then away from the highway, at right angles, 250 feet to the top of the mesa. But they went out *six* 50s instead of only five, so that "250 feet" became 300 feet. And that wasn't all! They didn't take the curve in the highway into account, so the rearmost 250 feet was short, and they had to fudge it out to a full 250 feet. There wasn't any acrimony, really. They even laughed about it until the movie star started building a residence on the high point. After the house was built they finally called on me to make a formal survey. It turned out that not only was the one side of the lot going away from the highway 50 feet longer than the other, but the other side was also messed up. So a good part of the house had been built outside the 250-foot square called for in their original deed. It was—and still is—an interesting house, and I did help them draw new property boundaries. But it would have been better if they had had the home site surveyed properly to begin with! Nobody ever wanted to pay what it was worth to straighten out royal messes like that!

"Locating" a Mine Through Clairvoyance

In 1952 I made a map of Yuma County as big as a table top. It got publicity in various mining magazines and led one of the sons of Edgar Cayce,[24] in Virginia, to contact me. His father had been a famous clairvoyant, but at the time I was contacted I had never heard of him. The son wanted to know if I could go out and find a mine for him in Yuma County. "What do you mean, find a mine?" He said he "knew" where there was one and gave me written directions that presumably came from his father. "Go to the King of Arizona mine at Kofa (which by that time was a ghost town[25]), and go in a direct line to the North Star Mine, and project that line, northerly, past the North Star (not the North Star itself but the North Star Mine), exactly 12 miles, and there's a possible gold mine at that point." It was at a point where at 4:00 in the afternoon on the 1st of each May a mesa would throw a shadow—right on the spot. He wanted me to determine where the "calls" that his late father's prediction had made would place him.

I wrote him that the whole thing just didn't sound right. In the first place there's a big mountain range to cross in that 12 miles—the Kofa Range. Also, there is a prominent and well-known topographical feature, New Water Pass, within a few miles of the supposed site. If a person wanted to describe a nearby

---

24. Edgar Cayce (1877-1945) believed his inspiration was derived from the person he was working with and what he termed the "Ashkanic record," an all-inclusive data base, representing all historical events.
25. I included information on Kofa in Volume 1.

site, he would surely, I felt, have begun his "calls" at a nearby commonly known place, intervisible with the possible bonanza, not from the 12-mile distant North Star mine. I was sure that one would be able to see this place, wherever it turned out to be, from New Water Pass. After all, it would be, from his description, only a few miles southeast of the pass.[26]

Yuma County desert with Kofa Mountains in background, 2005. It was in these mountains that Edgar Cayce "saw" a gold mine. Photograph by Robert Whitcomb.

The son sent me a paperback copy of the book *Edgar Cayce, the Sleeping Prophet* (Stearn, 1989) and said, "You can keep this." He said the mine had been located by clairvoyance. Maybe he didn't use that term, but that's what it amounted to. He said, "I don't want to spend a lot of money doing this." I said that it sounded interesting and that I would do it. I had a set of detailed U.S. General Land Office (GLO) maps of the North Star and the Kofa claims. I calculated a direct line connecting the principal shafts of the two mines and projected that line northwesterly 12 miles. In that part of Yuma County the GLO had run

26. I hadn't been within six to eight miles of the "subject" site before. (Yuma County was very large at the time; it was later split into two counties, Yuma and La Paz.)

out the township lines (6 miles square) but had not subdivided them, so there was a brass-capped government corner monument every half mile for 24 miles around each township's exterior. These monuments were set out by the GLO as reference points. This had been done maybe 10 or 15 years earlier. So I said, "I think I can find the corner nearest your site in a day or so and go on from there. Now I knew that all this was in the middle of nowhere! No one lived within 8 miles of the site, and there were no real roads. Nevertheless, I had grown up in Yuma County and wasn't afraid of the desert.

Meanwhile, I contacted a friend of mine who owned several mining claims near Patagonia and whose wife was a member of the Allis Chalmers family. He lived in Virginia Beach and knew the younger Cayce, who was in nearby Norfolk. I asked about Cayce, sort of checking up, and my friend said he was "A-okay." So I went ahead with it.

I was living in Patagonia at the time and considered paying someone to go with me. But the site was at least 300 miles away, and I wondered how I was going to keep expenses down. There was a friend of mine, Leland Wilson, whom I had known since he was a teenager in Tucson. He had a grocery store in Patagonia and wanted to get away from it for a while, so he said, "It sounds like fun. I'll go along with you and just not charge you anything."

We drove over to Yuma County in late October 1953. We found a deputy sheriff at Salome and told him, "We're going to be out by New Water Pass, and if we don't come back in a couple of days, send somebody out." This was standard procedure for me. I always told someone when I went out in the desert and even drew a sketch of my intended route and destination. You might be in good shape and able to walk out if your car broke down, but a broken leg or even a turned ankle could be fatal, even in cooler weather.

There weren't any real roads out there. All we had was primitive double-rutted trails. It took the whole first day just to find the first GLO monument. I had maps of the area, but they only showed lines 6 miles apart, without landmarks. I finally found a brass cap that was properly marked (the caps of the government monuments have numbers and letters stamped with steel dies). We camped overnight so we could get an early start the next day. Because it was getting close to November 1, which is six months from May 1, I figured that the sun would be positioned very close to the same part of the sky, so that the shadow would fall at about the same place at 4:00 in the afternoon. Sure enough, we got there at about 2:00, and the "12-mile" spot where the shadow was supposed to be was less than a half mile from the nearest brass cap of this line. I traced a line northwesterly by compass, pacing the bearing and distance I had calculated in my office, and, by counting strides,[27] we were able to find a marker every half mile—after we found one to start from, that is!

A big high-pressure cross-country natural gas line passed about a half mile to the north of the place Cayce had identified and continued on westerly through New Water Pass. A primitive road that the Natural Gas Company used for maintenance or inspection ran right along the gas line. The gas line was pretty straight, but because the topography was hilly, there were some pronounced zigs and zags in this service road. One part of the road made a prominent "figure 4" northeast from where Cayce's spot was. I drew a sketch of the landscape, so that if one got near the "target," that figure 4 in the road would show up plainly. The gas line made the vertical part of the figure 4, and the road weaving down the edge of the mesa formed the rest of the 4.

When I located "the spot" on the hillside, it was intriguing. There were quite a few round walnut-sized or hen's-egg-sized nuggets lying about. They were like large beach pebbles, made of solid hematite.[28] There was a nugget maybe every square yard, so in fact it *was* kind of a mineralized area! But I saw no rock outcrop. There was only "desert-varnished" gravel.[29] The hematite nuggets were the only evidence of anything unusual. But you'd have to dig down through the gravel to get to a "hard rock mine." It might be 50 feet or it might be 150 feet. *Quien sabe?* I reported back to Cayce, and he paid me immediately. But I never heard any more about it after that, and I think I would have if it had turned out to be a productive mine or a buried treasure. So that bit of clairvoyance didn't work; in that respect, it resembled my own experiences. None of my "hunches" ever turned into anything either.

Pvt. Bob Jones

I had another wild goose chase in the 1960s. Back in World War II days, an African American private named Robert Jones, who served at Fort Huachuca with the 25th Infantry, claimed to have found a stack of gold bars in a pit he had fallen into when hiking on Post land. His story was in the news. As the story

---

27. My strides (two paces) were close to the 5-foot military stride. In those days, with no GPS indicators or other equipment, we learned to take regular strides. Then we'd calibrate it, in feet.
28. Hematite is essentially iron oxide. It occurs in several forms and colors.
29. Desert-varnished gravel consists of rocks that have been polished over the centuries by daily sun baking.

went, when he found the cache, he marked his initials on a rock, but in reverse—"JB" instead of "BJ" for Bob Jones!

Years later, in 1959, long after he had left the Army, Jones returned to the base and requested permission to search for the treasure. In fact, he returned on several occasions.[30] I believe the time I became involved may have been his third visit. As I remember it, Jones hired an excavation outfit, which brought in a bulldozer and backhoe. He didn't have the money to finance the operation himself, but he had found somebody to back him. Well, the Army couldn't have these earth movers prowling around unrestricted on an Army reservation, so they asked me to lay out a square 400 or 500 feet on each side in the area that he wanted to search. The initials and the cache were supposed to be within the area that I surveyed. Jones's "sponsors" tried to show me the place, but what I was shown featured wooden "seepage-spring" covers with initials or names on them. I don't remember that any of them were marked with either BJ or JB. Anyway, I laid out the area that Jones's men thought was "hottest." They were supposed to restrict their activities to that area and try to restore it before they left. So I made a map and staked out the square on the ground, but I never heard any more of it. Like the wondrous gold mine Edgar Cayce "called," it became just another of the countless lost treasures of western lore.

---

30. Jones first returned to the base in January 1959. According to historical records at the Fort, a preliminary dig with shovels and bulldozer turned up nothing but water. However, the story was published in *Life* magazine. The publicity generated by this and other news accounts generated pressure for further searches. It was estimated that the cache described by Jones would have been worth more than 60 million (2004) dollars. In the fall of 1959, a more extensive search was conducted but also failed, and post authorities denied permission for further searches. However, in 1962, one of President Kennedy's military aides granted permission for further searches to be conducted, all of which were fruitless. A fourth failed exploration was made in 1968, a year before Jones died. In 1975 the Quest Exploration Corporation obtained permission for what was, at the time, the most comprehensive and scientific treasure hunt ever conducted. Physicists and engineers collaborated, using, among other techniques, microgravity sensors, which would have readily detected a cavity of the size Jones had described. Their tests failed to demonstrate the hypothetical hole in the ground. There have been no further attempts to find the treasure since that time (Anonymous, 2004b).

The Treacherous Canary in the Chiricahuas

Juanita Kirkendall grew up in the Patagonia area before the Depression. I met her parents when I was in Patagonia in 1929 on my field assignment. She married Ralph Morrow, a famous Arizona game warden, in about 1929, and they lived out the balance of their lives in the Chiricahuas. She told me and my survey crew a tale in 1956 or so, when we were surveying mining claims at their backwoods home.

The Morrows' residence above Hilltop, on the east side of the Chiricahuas, was isolated, but it was on a better-than-average U.S. Forest Service road and boasted a good deal more than the usual backwoods necessities. They had a Forest Service telephone on the wall. In those days, phones of this type were mostly on party lines, whereby each subscriber could summon "Central" (the operator) by cranking in a specific ring pattern—for instance, "three shorts" (i.e., three short, sharp twirls of the little hand crank on the side of the phone). Central would then send out a similar set of rings (for instance, "two shorts and a long") summoning the party that was being called to answer his or her phone.

Now each signal was received not only at the phone of the intended respondent but also at all the other six or eight homes on the same line, and, of course, everyone on the line knew exactly who was receiving a call. Some of them got to listening in on conversations that really were none of their business. That was a common pastime in those days wherever such party lines existed. There was an obvious disadvantage to this unacceptable practice—any tell-tale sounds from the "sleazer's" home could be heard by everyone else on the line. Therefore, the person listening in had to remember to eliminate any tell-tale noises. An easier way to accomplish the same purpose was simply to cover the mouthpiece of your own phone with your hand for as long as you kept your circuit open.

Ralph's work kept him away most of the daylight hours, and one quiet solitary day, Juanita heard Maw Smith's and the widow Brown's rings matching up. She admitted that she brazenly listened in. She had no appliances running or anything else creating noise at home, so she didn't bother covering her phone's mouthpiece. However, she forgot that she and Ralph had a pet canary—the only one on that side of the mountain. She confided to us that the conversation was getting to be quite interesting when one of the participants said to the other, "I don't recall ever having seen a canary at your house!" Juanita knew she was caught red-handed. She hung up the receiver (soundlessly, she admitted) and tiptoed back to work.

A Few Patagonia Legends

Any town that has been around for more than 100 years has legends, and of course there are many associated with Patagonia. I will recount a few with which I have had personal experience.

The Day Buck Blabon Wanted To Dynamite the Railroad Bridge

One of our local legends claims that there was a town elder who threatened to dynamite a railroad bridge. That one happens to be true. I know, because I was there!

In 1949, after a big rain, Sonoita Creek flooded, and water began to enter the town. There was water on McKeown Avenue, and there was even some flowing down Duquesne Avenue, although that may have been from canyons coming down off Red Mountain, rather than the creek, despite the dams the CCC had built in those canyons to prevent flooding. At that time there was a railroad trestle alongside the site where the highway bridge is today. The highway bridge then was a high steel-truss structure. The trestle, of course, was just a timbered trestle. It had a 210-foot span, as I remember. The waters were high and strong enough that they bent the trestle; although it didn't fail, after the flood it had become a big arc![31]

Well, the trestle, with its accumulated debris, was an impediment to free flow of the creek.[32] Water backing up behind the railroad trestle threatened to inundate much of Patagonia, and the town turned on its fire sirens to alert the population. By late afternoon—it was still light—the waters were cresting at such a high level that Buck Blabon decided he should take action. Buck was one of the most highly respected citizens in town and served by popular assent as an ad hoc "mayor" of the town, which had been incorporated only the previous year. He was a nice guy, honorable, and normally peaceable, but on rare occasions he showed a temper, and he was a man of action. On that fateful day, he brought a box of dynamite to the railroad trestle and said he would blow it up, if necessary,

31. Railroad workmen were able to realign it, and it stayed in service.
32. A similar flood occurred in 1983, 20 years after the railroad bridge had been dismantled. In that year, the highway bridge was blocked by brush and debris. The water went over the banks at the bridge and flowed through the site where the Post Office sits today and on down McKeown Avenue. I was out of town at the time and did not witness this flood.

to prevent flood waters from inundating the town. But Pemberton Blake, a section foreman on the railroad and another normally peaceable man, heard of Buck's plan and raced to the trestle to defend the railroad's property. The two men were yelling at one another and waving their arms, each determined to protect what was his. I stayed out of it—there was no need for anyone else to shout!

Meanwhile, however, other citizens were monitoring the rising waters. Just as the argument reached its peak, someone reported that the level had dropped half an inch. That might not sound like a lot, but in a flood situation it is huge. It means that the waters are receding. So the two friends were spared the necessity of choosing between the trestle and Patagonia.

The Portable Well

I had the opportunity, as a surveyor, to examine another, older Patagonia legend. Like so many of the figures who made early southwestern history, town founder R. R. Richardson was quite an "operator." Not long after he set up the Patagonia townsite, he arranged to have a well dug by hand on the mesa overlooking the town from the east, half on a lot of his and half on one belonging to a Mr. Wilson. It was to serve as a water supply not only for his own and Wilson's families, but for the entire town. The two men split the cost. Along about then, flaws were found in the 1900 Plat, one of which concerned the location of the westerly boundary of the townsite, which was also the northeasterly boundary of the San José de Sonoita Land Grant. Eventually, on March 23, 1914, Richardson recorded an "Amended Plat of the Townsite of Patagonia." By means of this, local folklore claims, the entire townsite was slid, "endo,"[33] along the railroad tracks about 4 feet. The legend does not specify in which direction the "slide" was made, but it was said to have put the well wholly onto Richardson's lot, thus neatly dissolving the partnership!

I first heard this tale from Charley Mapes and his wife in 1946, at about the time when Charley told me that he needed a lot survey. The Mapes residence was in Block G of the Patagonia townsite, and the Mapeses asked me to make a survey of their three lots. At that time I was just starting up my Patagonia business, and I wasn't planning to get into surveying, but I did have an excellent survey

33. "Endo" is a term describing the directionality of a shift. If a shift is in a direction roughly parallel to the axis of a rectangular or cylindrical component of the unit that is to shift, it is said to be slid "endo." I became familiar with the term, used in mining and general engineering, from my association with my father's railroad coworkers in my youth.

outfit, and I had had good experience in Bisbee, Yuma, and Morenci and sundry mining-claim conflicts. Also, I had surveyed and mapped gun positions along the shores of the Gulf of Mexico in the Army. So I said I would run out their lot lines if they could "sleuth up" a few original townsite survey monuments to start from.

Because Charley had been a railroad section foreman and had had frequent contact with survey crews, he knew what to look for, so he went about locating as many apparently reliable metal or concreted corners of town blocks as he could. He found three (the southeasternmost corner of Block I, the northwesternmost corner of Block G, and the southwesternmost corner of the north half of Block E). I checked out each of these three monuments and found that they agreed with each other as to relative location and as to the proper distance from the center of the main line of the railroad, which ran through the middle of town. I was determined to corroborate Charley's findings and found quite a few other "candidate" lot corner markers. All the corners pointed in the same direction. The evidence seemed irrefutable to me. I was satisfied that the lot lines had in fact "migrated."

Some months later, on May 17, 1947, I surveyed Lot 2, Block C, on the mesa overlooking town, for Mrs. Nellie Solano. Among other things, she wanted to know about the ownership of a well lying in the southwesternmost corner of her lot. The well had a stone curbing and was 57 feet in depth. It was a bit damp in the bottom (my tape came up with a little wet sand on the end of it) but had no standing water in it at the time. When I mentioned the survey to the Mapeses, Charley said, "Why, that's the well I told you about Richardson getting back by sliding the townsite!" And I believe it was, as nearly as I can piece the parts of the puzzle together. The stone collar of the well taped out to be 3.5 feet in diameter; its center lay 3.8 feet from the northerly sideline of Pittsburgh Avenue and 2.8 feet from the sideline common to Lots 2 and 4, of Block C. From this evidence, it would appear that the lateral "slide" or shift in the townsite was 2.8 feet toward Nogales (i.e., southwest). My estimate of 2.6 to 2.8 feet may or may not be in good accord with the legend, which claimed the lines had been moved about 4 feet. But it does seem that Richardson "slid" the lot lines several feet—to his own advantage, of course.

A Deadly Barrel

One tale, dating back to the days before there was a town, concerns the part of Sonoita Creek between Patagonia and Nogales. This section, while not encountering the problems that a major river would, has seen its share of disasters. The most famous was the flood in July 1929, which washed out the Patagonia-Nogales railroad. But there have been others. According to articles in the *Arizonian*, a Tubac weekly newspaper—from about 1859 or so—a rum runner came up from Sonora through what is now Santa Cruz County with a barrel of whiskey in an oxcart. There was a road following along Sonoita Creek to Fort Buchanan, but it didn't go along the creek at the narrows where Bohlinger Park and the Telles Shrine are today, because that narrow stretch of riverbank was wetland. Whether there was a beaver dam or a logjam (not likely) in the area we don't know. When the road was built they turned away from the creek and went uphill through a pass a half mile south of the bottleneck, as far to the east as Flux Canyon, about a half mile from its junction with Sonoita Creek. As the story goes, when the rum runner's group headed up the steep trail to the pass, the barrel rolled off the oxcart and crushed a young Indian woman in the party to death.

Today Sonoita Creek southwest of town hardly seems menacing. But, the area they were circumventing was not only marshy—it was malaria marshy. That problem was eventually solved or went away. Maybe as the water table dropped—as it surely has done in historic times—the marsh just dried up.[34] Also, Sonoita Creek is now channelized through the narrows. I would guess that was done when the railroad was built (1881-82). When the construction was underway the railroad no doubt built a road alongside the track to bring their teamsters in. That road evolved over the years and is now State Route 82, which connects Sonoita and Patagonia to Nogales. Every place that I've ever seen railroads there has been an accompanying road. Some have stayed primitive, but they always run right alongside the railroad, having been built—with teams in the old days—to afford day-by-day access for grading and construction crews.

Another Sonoita

The section of Sonoita Creek between Patagonia and Nogales is the site of another legend, which also seems to be true. A lot of accounts suggest that there was—briefly—a Sonoita other than the one we know today 13 miles northeast of Patagonia. This earlier settlement existed before the Civil War. It was about 3 miles below Patagonia, probably immediately northeast of the site of the old railroad quarry—where Bohlinger Park is located today. Walnut, cottonwood, and other trees were logged there. They had a sawmill, but I use the term loosely. I think it was just a saw "pit"—one man down 8 or 10 feet below pulling the saw down and one above pulling the saw up.

---

34. Fort Buchanan was situated near another series of malarial marshes on the Sonoita plateau.

Salero

Old mines generate legend after legend. A tale associated with the workings around Salero goes back to the days of the missionaries. "Salero" means "salt cellar." Supposedly, the bishop or some other high-level church official was coming to the Tumacacori mission 12 miles west of Salero, and the padres at Tumacacori said, "Look! We have planned a feast here, but we don't have a salt cellar!" They had two or three days advance notice, so they sent an Indian runner to a known silver mine 10 or 12 miles east. He came back with a chunk of silver ore, which they refined overnight and cast into the semblance of a salt cellar. I assume that it was just a sort of saucer from which you could pick up a pinch of salt in your fingers to season your food. I don't think it was anything fancy—I don't see how it could have been. They would have been doing well just to refine the ore so they could use the silver. In any case, says the legend, they were able to greet their guests royally.

MAPS

Maps, Maps, and More Maps

I have more than 10,000 maps, covering mostly southern Arizona and northern Sonora. My collection began when I acquired a large number of maps from my friend Hugo Miller, the Nogales assayer. I love them all, but some—those of historic significance—are of special interest. Let me share a few of these.

First a word about Hugo Miller. When I returned to Patagonia after World War II, I hadn't really intended to do survey work. However, my friend Charley Mapes talked me into doing "just this one job" for him—some lot maps of Patagonia. I asked Hugo for guidance in finding good, solid base points to work from. Hugo had been in business in Nogales since 1914. He leased his office from the SP. It was on Grand Avenue, alongside the railroad tracks, and in about 1949 the SP raised his rent. It made sense for us to share his office. I agreed to pay him a monthly fee for desk space in his office, which I kept until his death 15 years or so later. While I was there, he let me have almost daily use of his ancient map files. I augmented the collection later with more than a thousand maps from the collections of Nogalians Bracey Curtis, W. H. Roper, and other friends of mine. In addition, I was county engineer for several years and during that time made a set of "office" prints of many of the maps on file at the courthouse.

You can't be interested in maps without becoming interested in the history associated with them. Gradually, as I continued to add to my map collection, I became interested in the general history of the U.S. Southwest and northern Mexico. So I began to pick up both old and new maps (some dating as far back as the 1700s) and books on Mexico (several, in English, before 1830) and the American Southwest (originals from 1846 on). In all, I have probably close to 1,000 maps—originals, as well as copies—concerning early-day conditions. Over the years, I have also built up a collection of the basic narratives of exploration and development of the region.

Looking for Old Claims

One of my favorite maps is 127 years old. It was drawn with ink on cloth in June 1878 and bears the name of Solon M. Allis, a pioneer U.S. Deputy Mineral Surveyor. It shows many of the claims in the Patagonia Mining District, as well as the Aztec and Tyndall Districts. Over the years I have had a lot of fun going out to areas that were rich in claims and looking for them. I can still identify several of the early claims. Two of these, going from north to south, more or less, are the Dayton and the Helvetia. The Helvetia lode—like quite a few of the old mine sites—is now privately owned. Continuing southward from the Helvetia, there's the Tumacacori, the Lully, the Jefferson, and the Georgia. At Salero the map shows the old hacienda and the Mills lode, close to the ranch headquarters. Salero Peak still stands out, as do the stone ruins of the ghost town of Toltec. Actually, Toltec Camp and Toltec Mine were almost a mile apart. Coral Hill is on the map, too; it is called Squaw Peak now. The Missouri and the General Craig are still there. They're close to the county road, close to Toltec. The Empress of India was in that same vicinity.

Several years ago I visited the site of the Montezuma. Back in the Apache days, when miners were down in the shaft, they had to station a man on top of the hill to watch for the Indians, so they built a little fort for his protection. The last time I saw this, circa 1950, a rubble boulder breastwork was still there. There are other claims shown on the map. Over the years, I have visited 10 or 12 of them and have surveyed or mapped their workings. They include the Royal Blue, the Jefferson, and the Georgia. Over in Temporal Canyon is the New Discovery Mine. There's also the Leopard, the Essex, the Boston, the Saratoga, the Julia, and the Buffalo. All these should be on record at the courthouse in Tucson because they were patented before Santa Cruz County was split from Pima (in 1899). So of the hundred or so claims on the map, quite a few of them can still be found. Today, however, their value is in real estate.

The claims I mention here have all been surveyed for patent and have Arizona serial numbers. The last surveys of this type that I ran (in 1974) had serial numbers around 4,500. That indicates that there had been that many mineral surveys "run out" (surveyed) in all of Arizona.

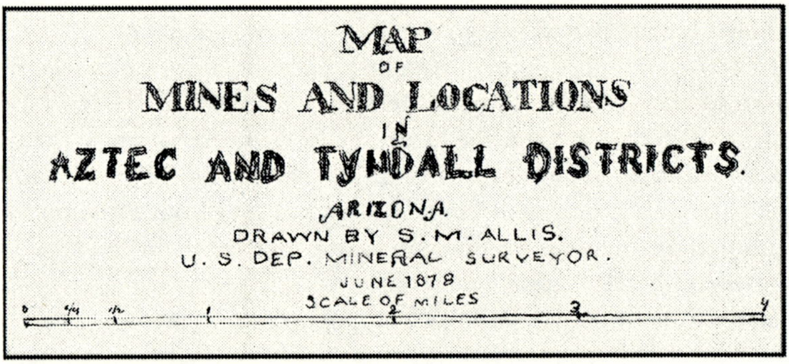

MAP
OF
MINES AND LOCATIONS
IN
AZTEC AND TYNDALL DISTRICTS.
ARIZONA.
DRAWN BY S. M. ALLIS.
U. S. DEP. MINERAL SURVEYOR.
JUNE 1878
SCALE OF MILES

Two details (facing page) from an 1878 map drawn by Solon Allis, an oldtime U.S. Deputy Mineral Surveyor, showing mines and locations in the Aztec and Tyndall Mining Districts. The top detail shows the Salero, Jefferson, and Georgia claims, among others. Note the location of "Old Selaro [sic] House." This house served as the mine headquarters, then as ranch headquarters. The lower detail shows the Empress of India, the Missouri, the General Craig, and the Montezuma. The road to Sonoita marked on the map corresponds to the current county road to Patagonia via the "old Sonoita" mill site along Sonoita Creek, just west of today's Flux Canyon.

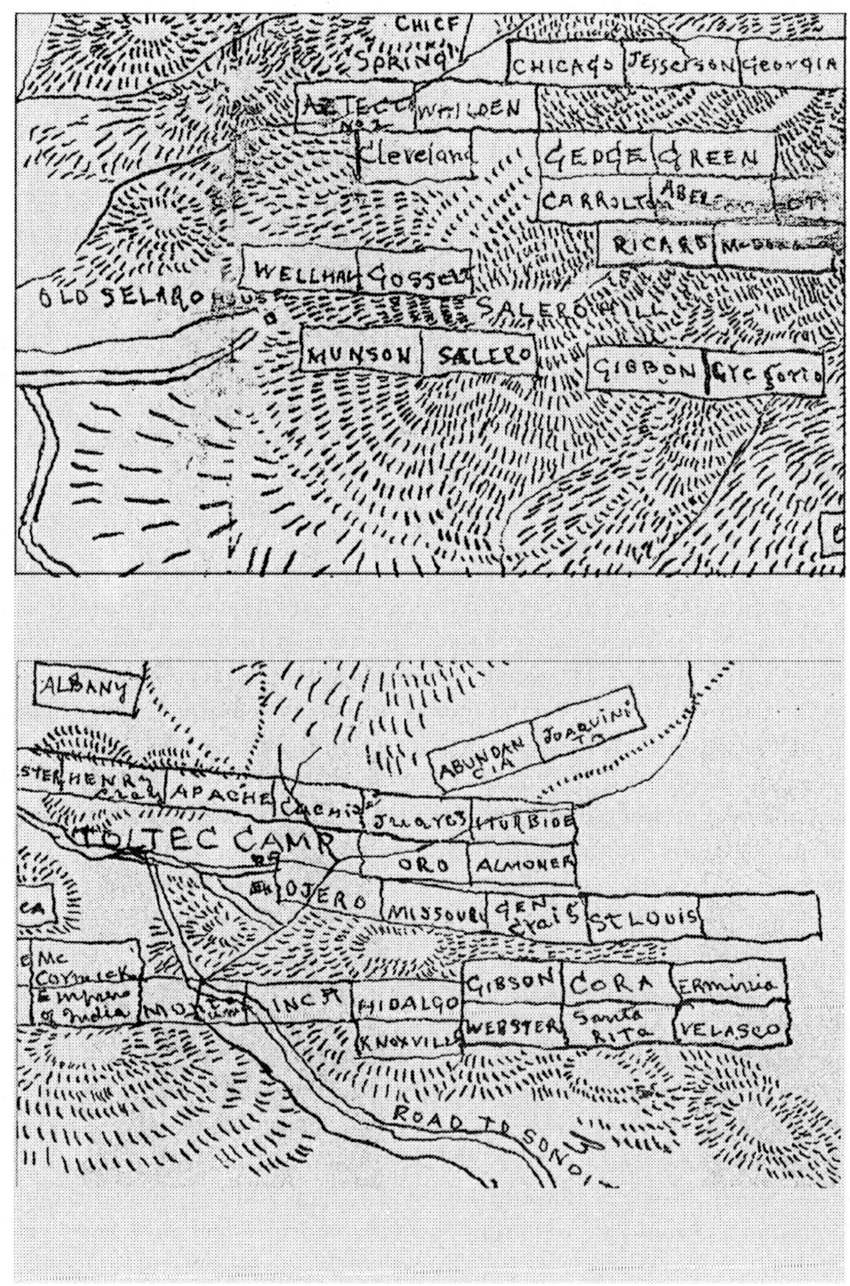
CHIEF
SPRING
CHICAGO
Jefferson
Georgia
AZTEC
No. 2
WHILDEN
Cleveland
GEDGE
GREEN
CARROLTON
ABEL
RICARD
WELLMAN
GOSSETT
OLD SELARO
SALERO HILL
MUNSON
SALERO
GIBBON
ALBANY
HENRY
Craig
APACHE
ABUNDANCIA
JOAQUINITA
JUAREZ
ITURBIDE
TOLTEC CAMP
ORO
ALMONER
OJERO
MISSOURI
GEN
Craig
ST LOUIS
Mc
Cormick
Empire
of India
INCA
HIDALGO
KNOXVILLE
GIBSON
CORA
Erminia
WEBSTER
Santa
Rita
VELASCO
ROAD TO

How does one find the site of an ancient claim? When I went out looking, I always looked first for a "claim corner." Each claim corner would have been marked with the initials of the claim name; for instance, the "Montezuma Fraction" would have an MF cut or stamped on it. MF 1, 2, 3, etc., in addition to the assigned survey number: 142 or 199 or whatever. The "accession" numbers of the surveys on my 1878 map are small—they're all less than a thousand. What I might hope to find at the site today would be the four or six corners of the claim. The corners were often marked on a field stone about the size of a big watermelon. Sometimes the marker, when it was intact, would be a squared timber post—3" by 3" or 4"by 4", with 3 feet or so showing and a pile of rocks helping to support the post. When you're looking for the old claims, you sometimes find the pile of rocks, but you wouldn't find any posts today. Those are long gone—as they said in the old days, "where the woodbine twineth and the cockaroo bird pineth for her first born!"

The World's Fair Mine

The Harshaw Mining District probably was the best known of the local districts. I have a map of the World's Fair Mine in that district that was made in 1893 by Gustavo Cox of Nogales. The mine was never surveyed for patent, but Cox's map shows the corner locations and distances pertinent to the survey. The original of this map was so brittle that it broke into pieces. I pieced it together, had the glued composite photographed, and made a few blueprints from the photograph. The owners of the mine were Frank and Josie Powers. I met them in 1930, when two friends and I visited Patagonia from Bisbee; one of those friends had met Frank Powers a few years earlier. The Powerses were elderly at the time of that meeting, but they were also quite sharp and spirited. The map shows their house and others, which were still there in 1930.

There are many tales about the Powerses. They didn't look or act like they had much money, but they did. The World's Fair was worked principally from a tunnel whose portal was (and still is) along the south bank of Alum Gulch. Back in the tunnel about 4 or 5 hundred feet was a winze[1] that they sunk on the vein. My map shows four claims in the Powerses' holdings—the Silver Lining, the World's Fair, the Mountain Chief, and the Josie. Although the mine had high-grade silver, the Powerses mined it in a desultory manner. They knew that they had a bonanza underground and could afford to shut down the mine, padlock the heavy steel gate

1. A winze is an underground shaft. Its collar is not out "in daylight" as a shaft's collar is.

at the portal of the tunnel, and take a trip around the world or go to New York City for a few months or whatever they wanted to do. Then they'd come back and reopen it, get another "stake" together, and away they'd go again.

The Powerses nurtured a running feud with R. R. Richardson—they could not stand the sight of him, and he felt the same way about them! Once, around 1900, the Powerses' crews were loading wagons with ore to send to Patagonia for shipment by rail to a smelter. After one weekend, quite probably a long holiday one, their men returned to work and found that, in their absence, someone had dug big (4' by 6' by 8') discovery shafts at two narrow places in the canyon where the mine's two "haul roads" ran. When one set of teamsters got to one hole they were so jammed up that they couldn't even turn their wagon around. They walked in to town and reported this. Another team headed out from town in another direction and hit the other pitfall. There were freshly prepared "claim papers" in stone monuments,[2] claiming both locations. It "just happened" that the holes had been dug on "showings" that had been lying, unnoticed, in the middle of the two roads.

Well, it didn't just happen, of course; it was intentional! So the Powerses said, "Who claims those discoveries?" "Richardson claims them both." So Frank Powers came storming into Patagonia to see Richardson. He said, "Well, I'm going to the Courthouse and see the records. What are the names of those claims?" RRR leaned back in his chair: "The Irish Lord and the Dago Queen." Powers, a blacksmith by trade, was Irish, and his wife was of Italian extraction. In those days, there was no such thing as political correctness, so those were the names given the two claims by Richardson, who had worked hard thinking up insulting nicknames for the Powerses and informing them of his choices. Even if there had been "PC" in those days, Richardson most likely would have done the same thing. But in the end, he couldn't legally block their roads and had to abandon his "nonmineral" claims. However, he had achieved his purpose.

---

2. As I described in Volume 1, the "location notice" was an important part of the process of staking a claim.

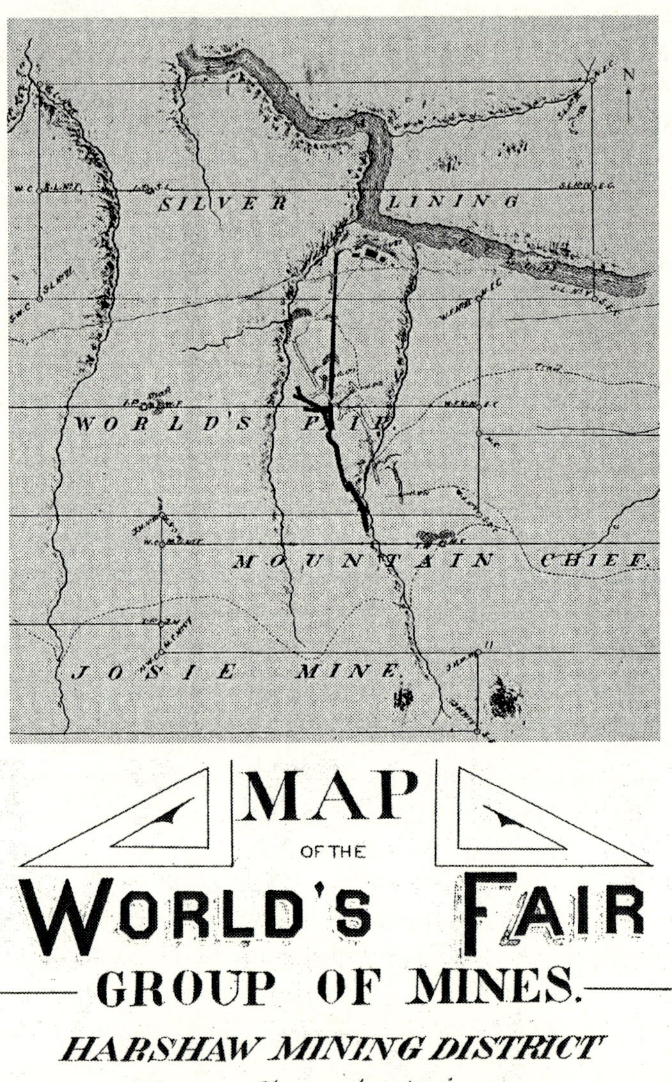

The World's Fair Group mines, owned by Frank and Josie Powers. The map detail above shows the Silver Lining, World's Fair, Mountain Chief, and Josie claims.

Mines Around Patagonia During World War II

On page 182 of *The Handbook to Arizona* by Richard J. Hinton, printed in 1878, there is a small fold-out map that shows the "French Mine."[3] This probably was a longhand error meant to be "Trench," inasmuch as the cursive F and the cursive T are so similar. I have a copy of that map. The Trench was given that name in the 1850s because of excavation holes left by Indians, who "trenched" the area for reddish body paint and pottery coloring. What was mined there depended on the current metal prices. Like most local mines, the Trench was at first a silver mine. But silver mines had to be abandoned after a while because at their deep levels they "turned to pyrite,"[4] besides accumulating lots of water. This was the early fate of the Trench, but it was revived and was worked during World War II. A local rancher, the late Norman Hale,[5] was master mechanic there. Having been born in 1914, he was old enough to serve and could have been drafted. However, he was deferred, and properly so, because of his expertise and experience, which enabled him to keep the mine going. Asarco had a little settlement up there, with maybe 30-35 houses. It was quite an operation. They had good water from a big reservoir off the old Hardshell Mine's shaft. There was no school; a bus picked up the kids and took them to Harshaw. When I got out of the Army I bought one of those houses, but I never lived in it. Instead, I rented it out and finally sold it. The buyers sawed it in half and moved it to town. It still survives as part of a bigger house on Pennsylvania Avenue near Fourth Avenue.[6]

Locating Old Graves

I have a map showing the sites of old graves in the Patagonia area. Arizona, with its Indians, outlaws, and sickness, was a dangerous place in the old days, and there wasn't always time or resources for a proper burial for the fallen. Christopher Columbus Watkins, a veteran of General Sherman's march to the sea in the Civil War, homesteaded near Patagonia in the early 1880s. The remnants

---

3. There is a French Mine in Arizona, but it is in La Paz County.
4. "Pyrite," actually iron pyrite, is a sulfurous mineral, and as this mineral accumulated the mine waters became increasingly acidic. The acidic water then tended to accumulate dissolved copper and other metals (in the form of ions). Sometimes the accumulated copper was sufficiently concentrated that it was actually valuable, but most of the time it was just a major nuisance.
5. Normal Hale passed away on December 25, 2004, at the age of 90.
6. Today, abandoned houses are seldom moved—they are demolished. In the old days, we didn't have such luxuries. My son and his wife live in a house in Tucson that they moved from one site to another.

of his house are still visible out Harshaw Road a mile from Patagonia. His wife and daughter are buried on a point of the mesa east of town; there's a nice memorial stone, surrounded by a fence, which the Watkins family has maintained for more than 100 years. The family was based in Bisbee for many years—even before 1900—and was well thought of there. Watkins's three sons were all miners and were familiar with all the country around Patagonia. In the early days they worked at the Total Wreck mine, in eastern Pima County close to Mountain View. The Watkins boys were friends of mine in Bisbee during 1930-40.

In 1894 a military officer came through the Patagonia area with an enlisted man or two and contracted with C. C. Watkins to exhume the bodies of troops buried in scattered, poorly marked graves in this area and bring them in for formal reburial in national cemeteries in California. Watkins's oldest son was 14 or 15 at the time and went along with his dad. Many years later, he told me about the operation. The officer would say, "Let's dig here. Looks like a grave." And Watkins would get $5, because that was what his contract called for. Even if it was a "dry hole," that's what he got—he filled the hole back in and got his $5. He was given another $5, for a total of $10, if he found a body. Davidson Canyon, which goes under Interstate 10 east of Tucson, was named for one of the men whose graves were found—a corporal who carried the Army mail from Tucson in a buggy. He was ambushed and killed by Apaches. As was common in those days, he was buried in a grave right alongside the road.

Watkins's son told me that they took the reburials to Fort MacArthur, which is part of the harbor defenses of Los Angeles. But Arthur Woodward, a noted Patagonia historian, said that they had all gone to the Presidio at San Francisco. In any event, the remains were gathered up in 1894. Young Watkins got 50 cents for making a casket to specifications for each body they found, and they were taken away for reburial. The boxes were more or less like orange crates, with two compartments, except that they were 3½ feet long. The compartment at the head end was about a foot long, about 6 inches high, and maybe 14 or 16 inches wide. The small end was big enough for a pelvis and a skull, and the other was long enough for leg bones. So the compartments were about 24 inches and 12 inches long. My friend got 50 cents per box for making them.[7]

---

7. A sewer line extension in a dedicated street in trackside Patagonia, which I laid out as town engineer in 1988, cut through several burial sites of unknown bodies. Because they were buried in boxes like the military ones, they probably were reburials. About 15 years ago, when they were extending sewer mains in the "East Side Subdivision" of Patagonia, we came across at least three bodies in a sewer ditch we were digging. The county officials decided to abandon the ditch and put the sewer on the other side of the alley, cover the graves back up, and record the details of the spot on a map. There is no marker there—the burial sites are documented only by a notation on my map in the County Recorder's office.

The "Lost" City of Calabasas

I have a map prepared by a skilled draftsman that shows the entire Santa Cruz region of the Arizona Territory, marked "1878" in pencil. It almost certainly was prepared to promote the "new" town of Calabasas, at the junction of Sonoita Creek and the Santa Cruz River. It shows "Calabasas," Tubac, Tucson, Arivaca, San Xavier Church, and the Santa Rita Mountains, as well as the Atascosas, the Sahuaritos, the Patagonia Mountains, and the Mowry Mine. It also shows proposed railways, Camp Crittenden and Casa Blanca. The map actually shows a town of "Sahuarit*o*" (sic), which is the proper name in Spanish—it comes from *sahuaro*, the cactus; a small *sahuaro* is a *sahuarito*—but today they call the town Sahuarit*a*. There was no Nogales or Patagonia then. Tucson appears as a tiny dot—it had only 2,000 people at that time! One thing the map doesn't show is the Santa Cruz county line—for a good reason; Santa Cruz County was not split off from Pima until 1899.

The map shows six GLO-surveyed townships in the vicinity of Calabasas, which was going to be the big town. The town did flourish for a little while. During the early years, the Benson-Nogales railroad, completed in 1882, passed through Calabasas, and there was a U.S. Customs office there also, some 10 miles or so north of the border.

Map of the Santa Cruz Valley region circa 1878. Above: title script. Next page: detail showing plans for "Calabasas," a conceptually large town that never quite became a reality. The map also shows the locations of the Tyndall and Aztec Mining Districts and Camp Crittenden. The grids of the Calabasas region correspond to 640 acres. The inscription "Tumacacori and Calabasas Ranchos" may have represented one of the grids. Tumacacori Mission and the town of Tubac are also shown on the map.

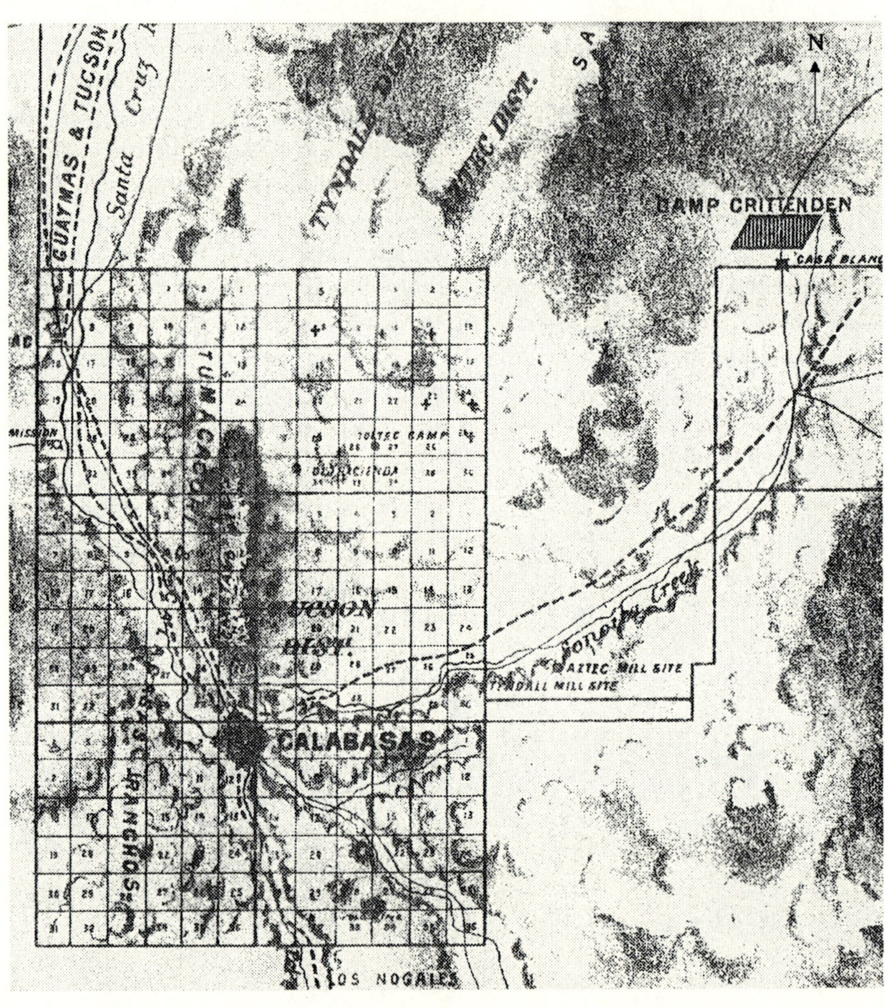
GUAYMAS & TUCSON
Santa Cruz
TYNDALL DIST.
AZTEC DIST.
CAMP CRITTENDEN
N
TOLTEC CAMP
OLD HACIENDA
TUMACACORI
AZTEC MILL SITE
TYNDALL MILL SITE
CALABASAS
RANCHOS
LOS NOGALES

The Calabasas area adjoined an old land grant (the Baca Float), whose status was long clouded by litigation. At the time Calabasas was "founded," this grant was not legally recognized. Nevertheless, people homesteaded there and bought and sold property. Finally, however, their titles were judged to be invalid, and everybody was evicted (in about 1914). So Calabasas didn't become a big deal—or even a little deal, unless you count a big hotel that was there for many years! The big town turned out to be the twin cities of Nogales, Arizona, and Nogales, Sonora. The beginning of the end can possibly be traced to the date that the Customs Service began collecting duties at the U.S. border at Nogales instead of at Calabasas. By 1950, Calabasas was just a siding on the railroad. Events like this remind us that we can make maps of land the way we think things are going to turn out. But reality often changes our original ideas and plans, and the end product often looks nothing like that which was originally imagined. The Calabasas townsite doesn't correspond to anything that we'd recognize today. It isn't even a ghost town. There is nothing that I know of at the "townsite" to commemorate the grandiose idea that once was.

The "Cities" Were Smaller Then

I have another map, stamped 1895 but with a copyright of 1898, that shows Santa Cruz County. I don't quite know how that came to be unless it was made by someone who could see into the future! The map gives population figures for Arizona towns, set out in even thousands: Tucson was 5,000, and Phoenix was—if you can believe it—3,000! Tombstone, Yuma, Prescott, and Bisbee were 2,000 apiece; Florence and Nogales were each 1,000. Those were the BIG cities! From then on down they're all in hundreds: Tempe 900; Clifton 600; Fairbank 500 (one-sixth the population of Phoenix!); Benson, Kingman, and Fort Huachuca 300 each; and Mesa 200. The population of Patagonia at the time of incorporation (on the 16th of February 1948) was about 500. I think it would be fair to say that Phoenix and Tucson have grown a bit faster than Patagonia! Oh, what a difference a century makes!

The town of Patagonia began to emerge in about 1898. Many people don't realize that towns start gradually. You can't say, for example, that the town started at 15 minutes past 12 on July 9. No town does it that way. When does it start? Is it at the occupancy of the fourth house or the 14th or the 24th? But maps from that era show that when Patagonia began to be recognized as an entity, Tumacacori and Tubac were well established. In general, the Santa Cruz valley opened up considerably earlier than the Sonoita Creek Valley. The Santa Cruz River is of course, much wider than Sonoita Creek, which is its tributary, and its valley is much, much wider. Hence, it was much more accessible to early settlers.

Chiricahua Apache Indians

In the old days many maps were constructed from military reconnaissances. I have a map showing the terrain of the Chiricahuas with respect to the military's battles with the Chiricahua Apache Indians. The maps shows telegraph lines. Also, however, it shows heliograph stations. Heliography involved the use of mirrors to transmit light signals, using sunlight or flame flashes in dot-dash codes. This technology dates from circa 400 B.C.,[8] when it was used by the Greeks, and is appropriate for rugged terrain. Although the map shows wagon roads, it also shows pack mule trails. Cochise County, in which this terrain was located, became a more populous county than Pima County when the Bisbee copper mines were in their heyday. So you see, times do change. The map itself, which was folded in the field and carried in mule packs, has no central detail because the terrain was so undeveloped. The title of the map and the legend are, in a way, more interesting than the map itself.

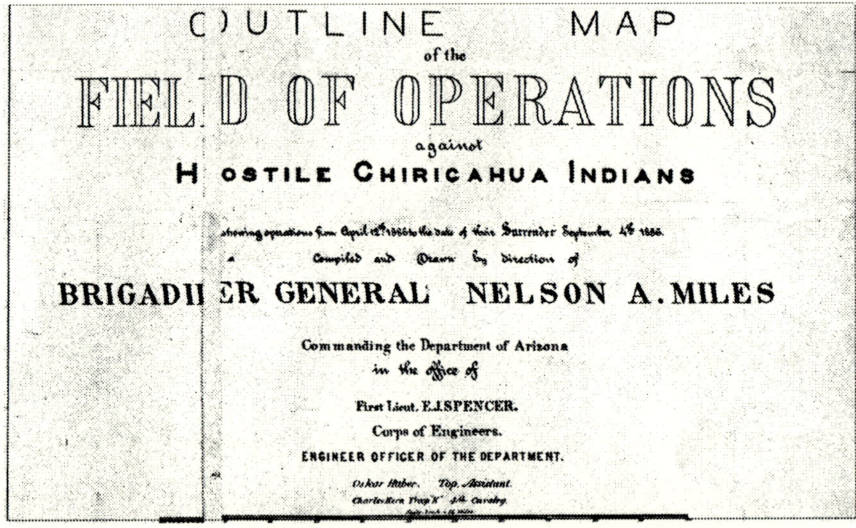

Legend for an outline map of the "Field of Operations Against Hostile Chiricahua Indians." Above: title of the map. Note the vertical fold line; the original folded maps were carried in saddle packs. Details of the map itself are less interesting than the legend. Facing page: legend details, including posts, stations, trails, roads, and boundaries.

---

8. Heliography is the art of communicating with mirror flashes in code.

LEGEND

Department Headquarters. (temporarily.)
Military Posts (garrisoned).
Military Posts (ungarrisoned)
Camp for Scouting & Observation.
Heliograph Stations
Heliograph communications
Telegraph Lines.
Indian fights.—Indian
Rail Roads.
Creek (dry)

Wagon Roads and Trails practicable for Wagons.
Trails practicable for Packmules.
International Boundary.
States & territorial Boundaries and Boundaries of Indian & Military Reservations.
(In U.S.) Limits of Districts of Observation under command of senior Officers therein.
(In Mexico) Prefecture Limits.

Border Trails

Today's atlases of the United States show improved and unimproved roads, interstate highways, roadside parks, campgrounds, historical markers, cities, big towns, little towns, and sites that were once significant but that are ghost towns today. It is not easy to imagine maps that don't show any of those features. But once upon a time none of those features existed! I have a map dated 1847, made by W. H. Emory, Lt. USA, that shows old trade routes, civilian trails, migratory trails, and army trails. The route of a pack train in 1846 is shown. The Mormon Battalion[9] led by Lt. Col. Philip St. George Cooke followed—for a way—a route very close to the present-day U.S.–Mexican border. There was a famous wagon road—the "Leach Road"—made for stagecoaches but little used; it goes along the San Pedro River from today's Benson and past the Casa Grande Ruins, bypassing Tucson and going on west past Yuma.

---

9. The Mormon Battalion (Ricketts, 1996) consisted of 600 members of the Church of Jesus Christ of Latter-day Saints who had been recruited by the U.S. Army in Iowa, on orders of President Polk, to participate in a campaign whose purpose was to take California from the Mexicans. The Mormon church had been founded in New York in the late 1820's by Joseph Smith and had begun to move west in search of a homeland where Mormons could practice their religion. Because of rising tension between the United States and Mexico over claims to Texas and the Southwest, Polk considered the Mormons a threat to the continued westward expansion of the nation; given the history of their persecution in the eastern United States, the Mormons appeared to be potentially hostile to the Union. However, Brigham Young, who had succeeded Smith as head of the church after the latter's murder in 1844, sent letters to members of Congress to affirm the loyalty of the church to the United States. For the Mormons, the matter was not just a matter of loyalty but of self-interest—they needed resources to accomplish their planned emigration and realized that their loyalty to the Union might gain them government patronage for their westward migration.

   Led by Lt. Col. Philip Cooke, the battalion—600 men and a few women and children—left Iowa in July 1846, on what became a fearsome journey (California Parks, 2003). Of the 37 wagons that started out, only eight reached San Diego. The expedition followed a route across the southwestern deserts that had been traveled by early explorers such as Father Kino and Juan Bautista de Anza. The settlers finally arrived in Palm Springs exhausted, hungry, and thirsty, probably surviving in great part because they had crossed the region in January, when the weather was relatively cool. On January 29, 1847, the group, including four women and a child, reached San Diego.

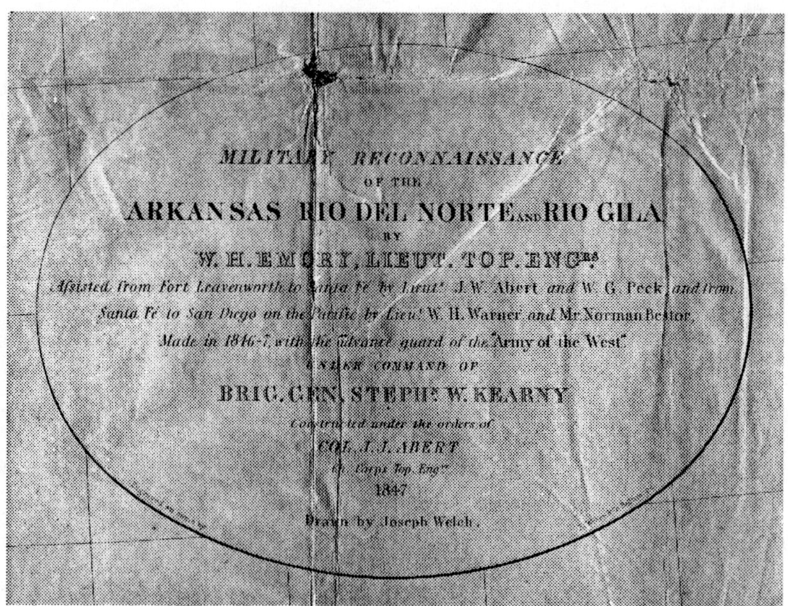

Military Reconnaissance of the Arkansas, Rio del Norte, and Rio Gila map title and information on the engineers and army officers involved in making it.

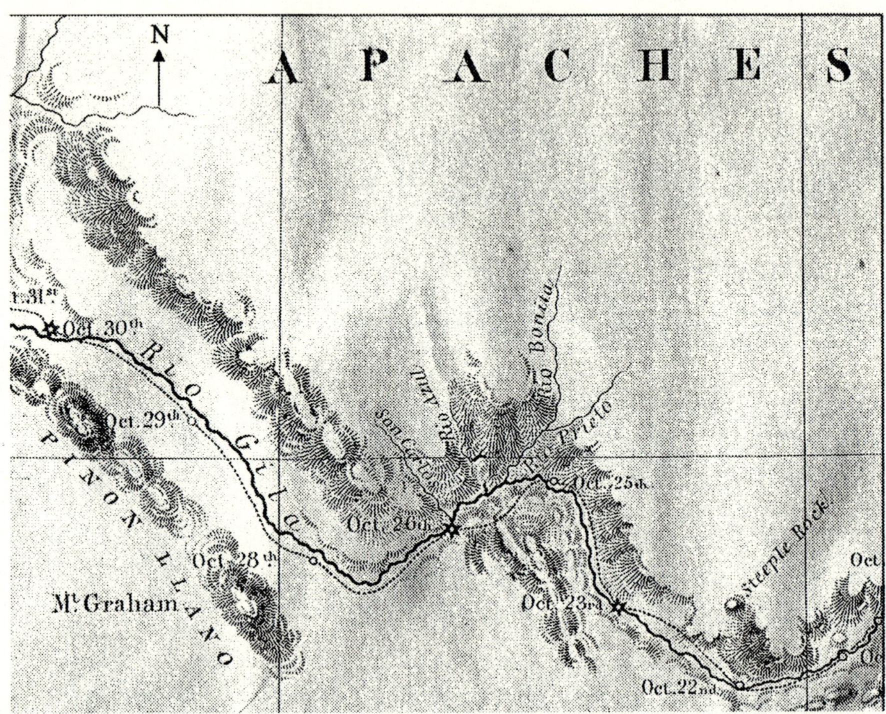

Details from Lieutenant Emory's military reconnaissance map showing various trails across the Southwest in the years before the Mexican war. Emory served under Brig. Gen. Stephen W. Kearny, who was sent with a small brigade to capture California. The detail on the facing page shows the route taken by the Mormon Battalion under command of Lt. Col. Philip Cooke. This route had been used by earlier explorers, but Cooke's battalion was the first to take wagons across it. Detail above shows the upper Gila River north and west of Mount Graham. The route along this river corresponds to today's U.S. Route 70. The positions of Kearny's party on several days in October appear on the map. In those days, there were Apaches north of the Gila. This is also true today. The 1.6-million-acre Fort Apache Indian Reservation lies to the north and west of the land shown in this detail. The route that is now I-10, following essentially the course of the old Bankhead Highway (Weingroff, 2004), is designated on the map as a potentially good route "providing that water can be found."

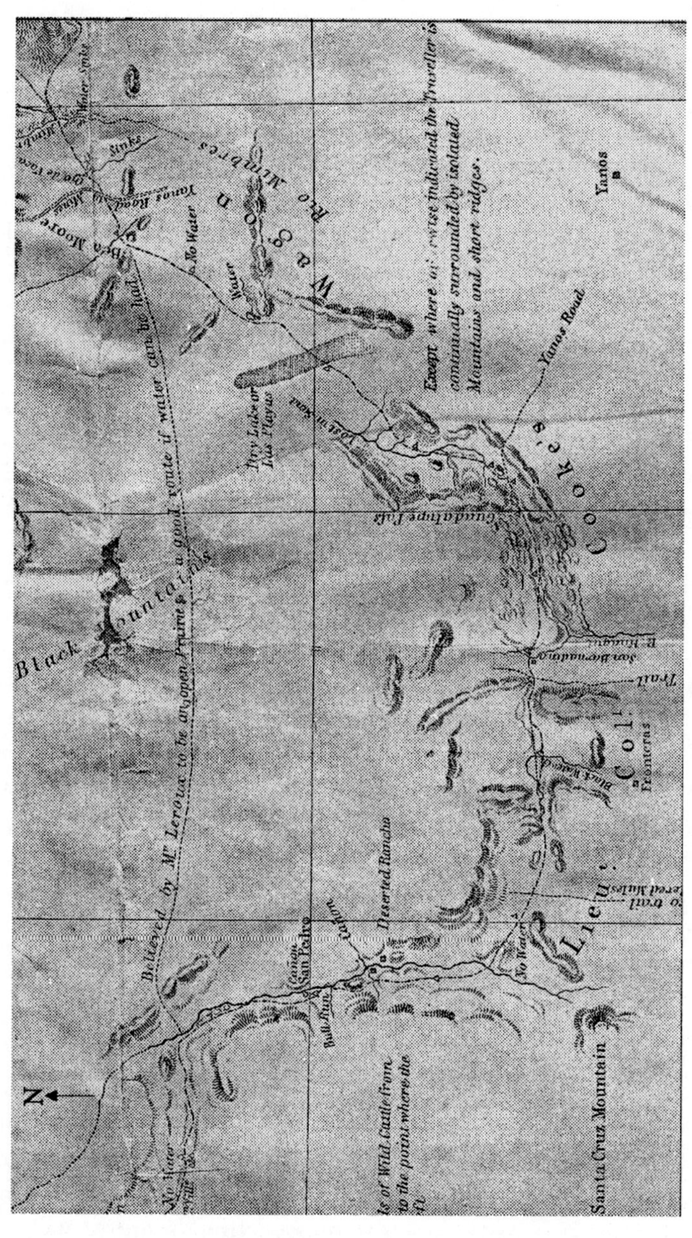
Black Mountains
Believed by Mr. Leroux to be an open Prairie & a good route if water can be had
Dry Lake or Las Playas
Except where otherwise indicated the Traveller is continually surrounded by isolated Mountains and short ridges.
Yanos
Yanos Road
Deserted Rancho
San Pedro
No Water
Fronteras
Santa Cruz Mountain
N

Ruins

I have a GLO map (not shown) that shows Fort Buchanan and Camp Crittenden after their abandonment. Before the Civil War, Fort Buchanan had a million dollars worth of supplies on site. In those days, that was big money! The intrinsic value was actually much less—a third of a million, say—but it had cost two-thirds of a million to get the supplies there from the east. All that materiel probably reached Arizona via Mississippi River ports. And in the end, the Union burned it all so the Confederates couldn't take it when they got there! As it happened, the Confederates didn't get much west of Tucson. Wars were expensive, even then!

The map shows the "ruins of Fort Buchanan," as well as nearby Warm Springs (today's Cottonwood Springs on the Rail X ranch) and Camp Crittenden. The adobe walls of Buchanan's post buildings were waist to shoulder high as late as the 1920s. When building the railroad through here circa 1882, it is said, some workers put tarpaulins over the tops of the old walls at Crittenden to make shelters. Even if they had just a corner of an adobe building to form a framework for a tent, they could make a structure that would shelter them from inclement weather. So even ruins can have practical value! A group of members of the Arizona Historical Society held a picnic dinner at Fort Buchanan in 1961, on the 100th anniversary of the fort's abandonment. Naomi and I attended that gathering.

On the GLO map a path along the Colorado River to the Rio "Buena Esperanza"—an early name for the Colorado—is marked. The path was called the Tizón. A *tizón*, a Spanish name for a "brand" or burning stick, was a small burning branch that Indians, traveling afoot, could carry along to warm their hands and to start a fire at the next camp. The Indians could sometimes be trailed by the discarded, burnt-out "stubs" left behind with their footprints along their trails.

So you see, every map is a textbook not only of geography but also of history. That is why I collect all these old maps. They are tickets to a past that is otherwise impossible to visualize.

Building the Obelisks

During World War I and for 10 or 15 years thereafter, Walter Riley, a real fine pioneer gentleman, was the treasurer of Yuma County. Arizona law at that time stipulated that the treasurer's job could be occupied by an incumbent for no more than two consecutive four-year terms. Walter was a Mason, as was Dave Inman, his principal assistant. The two men were good friends and very competent fellows. So that they could continue to serve the community, they would swap jobs every eight years—one would be the treasurer and the other would be the deputy. Then they would switch.

In the late 1800s, Walter had worked for the International Boundary Commission when they set up the obelisk monuments on the U.S.–Mexican border from El Paso to San Diego. These obelisks are all 7 or 8 feet high and probably weigh half a ton each. They are proportioned like the Washington Monument. The boundary numbers start with "1" near El Paso and increase as one goes westward.[10] Numbers 121 and 122 are in downtown Nogales. They were set at intervals of 2.5 miles or less so that a person walking along the border is never out of sight of the two monuments nearest to him—one in each direction. As surveyors, we speak of them as being "intervisible." Some of the obelisks are made of masonry, but most are made of cast iron. A few of the latter are cast in pieces for easier handling; they fit snugly together and lock. I've seen a picture of men hoisting the pieces up a steep cliff face with ropes and assembling them in place. They were cast in slabs, maybe half an inch thick. Walter worked on the obelisk job for a year or more when he was in his twenties. I don't remember him saying that he did really heavy physical work. He might have just driven a team. They had 10 or 20 wagons hauling the monuments, as well as tools, camp equipment, water, and rations. It was a big operation. After they were built, the monuments were repainted—white—every 10 years. Walter told us that he was awarded the painting contract one or two times in the early 1900s.

I have the government reports on that project. A set of maps is bound in a book that is 30 inches wide, about 21 inches high, and an inch and a quarter thick—I've never seen a bigger one. Its title is *Mexican Boundary Survey Commission Atlas, 1898.* I keep it in a special slot alongside a big bookcase. The maps show the topography and the "culture"—that is, such man-made features as buildings, fields, bridges, power lines, and pipelines—within 2½ miles on each side of the international boundary line.

---

10. The U.S.–Mexican border east of El Paso is the Rio Grande, which, given the shifting courses of rivers, couldn't be surveyed reliably even if one wanted to do it.

The author examining a map in the border survey book. Photograph by Robert Whitcomb, 2004.

Walter often told my Dad and me how the project was organized and what a major job it was. In 1857 survey parties set out from Nogales and Yuma to outline the southern boundary of the Gadsden Purchase. Their positions were determined by the positions of the stars. The people coming from Nogales were told to run due west as far as 111°00'00" west longitude, but they actually went to about 110°55'30" west (within about 4.5 minutes of arc, which is about 4½ miles). It had been agreed that the "accepted" results of the survey were to be final. So when the line that they determined in the survey was accepted, the border was fixed, minor mistakes and all. At Nogales, they were about 450 feet south of where they thought they were. They were supposed to be exactly at 31°20' north, and they set the specified "angle point" on that line at 110°55'30", to connect with the point on the Yuma end to be set on the left (east) bank of the Colorado River, "20 English miles"[11] south of the junction of the Colorado and the Gila River. The angle point is some 350 miles from El Paso, and the azimuth line runs 230 miles farther to the northwest. The line from El Paso to the angle point is a long, "easy," curve along the 32°20' parallel. You can see this "by eye" if four or five monuments are in your line of vision. The azimuth line is a direct (straight) line connecting the angle point and the southern end of the "20-English-mile" north/south line, on the east bank of the Colorado.

When they determined the coordinates for the Colorado-Gila River junction, it was under water (those were the old days when the Gila River still existed as such). But to surveyors, something like this is no problem. A man goes out in a rowboat, cord in hand. With the other end of the cord, the head chainman sets a stake and ties a knot. The measured length of the string becomes the first footage southward and is then used as the first few feet of the required 20 miles. As they say, "Close enough for government work."[12]

Determination of the coordinates for the border survey, as in all projects of that scope, depended on astronomical observations. The accuracy of the surveyors' observations depended on their knowing the exact time, as well as star positions, because, of course, the positions change with time. Watches in those days might be off by a minute or two in a year's time; hence, they used a "bank" of three many-jeweled watches and recorded the "rates" of each. In this approach, the first watch whose rate varied could be identified and its results discarded (two watches in close agreement serve to point out the "ailing" one). For that project they had the best watches available. They decreased the likelihood of serious error. The surveyors were pretty close, given the technology in those days. Of course today's GPS (Global Positioning System) equipment takes you to within no more than a few feet of a particular geographic position.

The surveyors chained or triangulated the required "20 miles south" and set a second angle point south of Yuma near what is now called San Luis RC. There are

---

11. The English mile is the mile that is universally used in the United States—that is, 5,280 feet or 1,609 meters. The odd number 5,280 derived from a surveyors' measurement of 80 chains of 66 feet each (Gunter's chain of 100 links created a portable length standard that was less stretchy than a cord). Other types of mile are the Roman mile (1,479 meters), the metric mile (1,500 meters), the nautical mile (1,852 meters) the "geographical" mile (7,420 meters), and the Prussian mile (7,532 meters) (Calvert, 2004).
12. Today they would have more trouble with the location of the river junction than they did then. Now there is no junction. The Gila River has been diverted or, rather, extirpated, by a series of upstream dams (McNamee, 1994).

many San Luises in Mexico. This one is "San Luis, Rio Colorado" because it's on the west bank of the river. At the time of the survey, there was nothing there at all.[13] The eastbound survey party was led by Lt. Nathaniel Michler. Starting from the determined point, they headed southeast across the desert. They had to plan carefully to conserve water. They had several ranges of mountains to cross, including the Tinajas Altas range, 25 or 30 miles east of the Colorado River. These are the mountains with natural water holes that I described in Volume 1. That water saved the life of many an animal or man, but it wasn't nearly plentiful enough to subsidize a large survey party, with its animal-drawn wagons. Also, the *tinajas* were several miles north of the line they were trying to establish.

The surveyors started off in June or July with their teams (probably four-horse teams) carrying four wooden barrels of water per wagon. They would go as far as they could (perhaps 20 miles), stop, and let the horses drink the water from one of the barrels. The party would then leave a full barrel there for the return trip and take the other two full barrels to the turnaround point. They would plan the next step carefully to be sure they had enough water as they ran the survey beyond the first reach. It might take them two weeks or so to haul the water that far in big enough quantities to do the job.

So the two parties moved out, one from the east (heading northwesterly from Nogales) and the other from the west, each "chaining" along its respective lines (and probably setting temporary stakes at each mile for future reference) on what is called a "random line."[14] They were hoping their observations for latitudes and longitudes were accurate enough that they would be able to meet, way out there in the desert. But it proved to be too difficult to work in the desert in the dead of summer, when ambient temperatures could often be as high as 130°F.[15] Michler finally conceded that they would have to abandon the survey until cooler weather. Horses could do a lot of things in those days—people don't give them credit for all they could do. But they couldn't do the border survey in that desert sand and heat! It would be a big job even now, with air-conditioned trucks, GPS equipment, and everything else that modern-day surveyors have.

---

13. Now there's a sizable town on both sides of the border—particularly the Mexican side. The population of San Luis is more than 200,000. The valley there is flat, so there is farming on both sides of the river. On the U.S. side, the land is irrigated by "shirt-tail leavings" of the Yuma Project's canals.
14. "Random lines" are actually not random, but their temporary locations (as points of reference) are; the surveyors seriously use all methods at their command and attempt to approximate the actual line as closely as they can. However, these lines are temporary—they are "drafts" of a final line to be established in a more exacting survey.
15. Ground temperatures could be as high as 160°F in the dunes west of Yuma. I once (circa 1930) got a reading of 162°F on a scientific thermometer placed upright in the sand in that area.

NEW ROOTS

Visiting Dad for the Last Time

Early in 1946, shortly after I got out of the Army, I traveled by train to Omaha to see Dad, who was undergoing treatment for leukemia. The train stopped for a while—maybe 15 or 20 minutes—in North Platte. I discovered later that a young lady named Naomi Wagner had been volunteering there at the World War II welcoming service canteen at the time. I didn't know I was supposed to find her, so I didn't look! I got to Omaha and found my parents in reasonable health, considering Dad's leukemia. I stayed with them for three weeks or so. As I mentioned in Volume 1, when I went off to join the Army, I left my 1938 Chrysler with Mom and Dad in Yuma, and in 1943 they drove east in it. Within a few weeks of my arrival in Omaha, they had ordered a new Plymouth coupe. So I left Omaha driving my old Chrysler. I headed south first, to visit relatives in Salina, Kansas, and then drove on west to Arizona.

Mother, having been a nurse, took pretty good care of Dad, and his doctor was excellent. Eventually, however, he succumbed to his leukemia—as everyone did in those days. At the end, I flew to Omaha in time for a brief visit with him before he passed away. After his death, Mom and I drove to Grand Island, where Dad was buried in the Shoemaker plot.

My father, Harry W. Lenon (lower right), with three other C&NW "rails," circa 1940. The two men in the top row were brakemen. An old conductor, lower left, carries the bearing bestowed by his job's importance. Photographer unknown.

For quite a while several Shoemaker relatives had lived in the village of Burwell (population 600-800), out in the Nebraska Sand Hills. One of my Shoemaker first cousins had inherited a small house there. In those days, it was customary for families to take care of their own, so the cousin arranged for Mom to live in the house with two of her older brothers—one a bachelor and the other a widower. I had hoped she would come with me to Patagonia, where I had settled in. I could have found a good solid house for her. But, unlike Dad, she was willing to cast her lot with Nebraska—harsh winters and all. After all, she was familiar with Burwell, and she didn't know anything at all about Patagonia.[1] In Volume 1, I described some of the trials and tribulations that Mom had had trying to learn how to drive. But she never learned! So she was unable to come down to Patagonia to visit. So when she was settled, I drove their 1946 Plymouth on down to Patagonia and, as a result, had an extra car.

A Ranch Girl From Nebraska

It happened that Mom and Dad knew well—and favorably—a couple named Wagner, who were longtime Burwell residents. They had lived next door to my former surveyor uncle, Al Shoemaker. (The Wagners also had an indirect connection with our family through a male relative who had married one of my many Shoemaker cousins.) They had a daughter, Naomi, who was a bit more than 9 years younger than I, but my folks didn't know her, as she was then "away" teaching school in Grand Island. In the course of a letter telling me that Mother was settling in, my cousin—Mother's "landlady"—mentioned that Naomi was about to take a teaching job in Phoenix and that I'd better get up there and meet her. At the time I had a mining job going near Congress Junction, which is 70 miles north of Phoenix, so I was able to stop by and exchange Nebraska reminiscences with Naomi on my trips between Patagonia and Congress Junction.

At the time I started visiting her in late 1946, Naomi and three other teachers had rented a hard-to-find apartment in downtown Phoenix. So when I visited there, I actually met a whole houseful of girls—but I preferred Naomi. When I finally ventured so far as to offer to take her out for supper, the vote was 4 to 0 to have me come and eat with all of them! They had a neat arrangement that worked very well. They had divided the jobs of head cook, cook's helper, and housekeeper among themselves, leaving one girl unassigned. Each week, the four were "promoted" a notch in rotation. The four kept in touch for years. Two have passed away now, but Naomi and Juynema Prentice Steele, now widowed, are still close

1. She never did get to see the town where I chose to spend the rest of my life.

friends. Juynema (pronounced wy-nee-ma) now divides her time between Phoenix and Sierra Vista and stops by Patagonia every now and again.

Juynema was originally from Kansas City but moved to Grand Island, where she and Naomi taught grammar school. They eventually decided—as Dad had done—to "Go West" together and had found teaching jobs in Phoenix. Juynema's parents gave them a Ford coupe, which they called "Bessie." Bessie was at best "middle-aged," but probably was more properly considered to be elderly. She had some severe limitations, including being a bit of an "oil hog" (60-70 miles per pint of engine oil)! Juynema's parents were aware of this problem and knew that Bessie would need a lot of oil to get all the way to Arizona from Nebraska. They were accustomed to drinking a pint of prune juice at breakfast every day. So they saved the glass juice bottles and filled them with oil as a going away present for the two adventurers. The girls were cautioned to make regular "oil stops" to keep Bessie's digestion in proper health. It worked! Bessie not only survived the trip from Nebraska but served Juynema well for extensive sightseeing in Arizona during most of the five years the roommates taught together. Naomi was the last of the four Arizona immigrants to marry.

Naomi and longtime friend Juynema Prentice Steele, with whom she taught school in Nebraska and later in Phoenix, Arizona. Photograph (2004) by Robert Whitcomb.

Because Naomi's family and my mother were in Burwell, that's where we were married. On August 14, 1951, Naomi and I began our new adventure. Most of

Naomi's Phoenix friends couldn't find Patagonia on the map and at first seriously wondered about the wisdom of her move.[2] Nonetheless, she did move to Patagonia, as my bride. We rented a newly completed home in Patagonia for $40 a month, and the local folk arranged quite a reception for the bride in the form of an outdoor potluck supper in Richardson Park. We lived in the rented house for the first five years of our marriage. Naomi worked in the office, keeping the files up to date, attending to the billings, and doing some drafting. She had had some experience in drafting as a summer employee in the Nebraska offices of the U.S. Soil Conservation Service. And she still takes excellent care of our business transactions, here, in 2005!

Naomi Wagner Lenon, 2004. Photograph by Robert Whitcomb.

We have lived in our current house on Pittsburgh Avenue since building it in 1960. We have a grand view of the town and of Mount Wrightson. I still enjoy looking down at the town's lights at night. The town has come a long way since

2. With the advent of better roads and better automobiles, Patagonia is now on the regional map of Phoenix travelers.

Dawson Scoggin's single streetlight, not so much by adding a lot of people as by continuing to improve the quality of life of the residents. It is singular that the town has been able to modernize without significant growth. And if we sometimes wonder if we are on the right track, we can look up at Mount Wrightson ("Old Baldy," el. 9,450 feet) for inspiration.

The Kids

Our first child, "Young Bob" (Robert Lenon III), was born in Burwell on October 31, 1952, while his Lenon grandmother was still alive. However, he had traversed six states by the time he was 6 weeks old, when he became an Arizona resident. He has been here ever since. At the time of his birth I was working on a mining job in Bisbee for Phelps Dodge. I was staking mining claims in large groups in preparation for exploratory drilling. I ranged around a good deal on similar jobs over the years. Bob and his wife, Nancie, live in Tucson. Their grown daughter, Eraine, recently moved to Texas. By the end of 1957, in addition to Bob, we had two daughters, Jeannie and Janet, both born in Patagonia, so it was a busy time. Our kids all had the pleasure of growing up in Patagonia. Jeannie and her husband, Keith Ecker, live in Indiana; they have a 10-year-old daughter, Kerrie. Janet is a National Park Service employee. For several years she has been stationed at Tonto National Monument near Globe, Arizona. Because this is just a five-hour drive north of Patagonia, she is able to visit us quite often.

About the Author

Robert Lenon was born in Nebraska but grew up in Yuma, Arizona. Influenced by old prospectors and miners he knew as a boy, he decided to become a mining engineer. He graduated from the University of Arizona in 1930. He then worked briefly at the Morning Glory Mine near Mowry, in the Patagonia mining district of Santa Cruz County, Arizona, before accepting a job as an engineer with Calumet and Arizona, a large mining company in Bisbee. As the Great Depression came on, C&A merged with Phelps Dodge, and he was one of many C&A employees who were laid off. He spent the bleakest Depression years scratching out a living in the Yuma Desert and working on the All-American Canal, a project in which Colorado River water was diverted to California's Imperial Valley and on to San Diego. After three years away from Bisbee, he returned to work for Phelps Dodge. However, after a while, he became restless under large company management and left to become his own boss. He operated a small but profitable tungsten mine in the Huachuca Mountains for a time and later supervised a gold mine near San Diego.

In 1941, when World War II was looming, Bob enlisted in the Army. He was first assigned to the Coastal Defenses of California and Florida and later served with the Corps of Engineers in England and France. When he got out of the Army in 1946, he opened a business in Patagonia, where he had spent two summers doing mapping field work around Mowry as a college student. When he began, most of the local mines had shut down, but he was determined to make Patagonia his home and there he stayed. With the help of Naomi Wagner, whom he married in 1951, he operated a firm that surveyed mines, drew up lode claims, and did general surveying work in a wide area in southern Arizona.

Throughout his career, Bob recorded stories and collected folklore and historical data. He was a member of the Board of the Arizona Historical Society for nine years and wrote several papers on Arizona history. His most recent paper was presented to a meeting of the Arizona Historical Society in 2001. In 1998, during the Patagonia Centennial, he contributed many hours of taped recollections. Those recollections, plus many of his writings, provided the material from which this two-volume set of books was derived.

APPENDIX I

Concentrating Ores (in Water) by Means of Differences of Gravity

Minerals vary greatly in their specific gravities (the amount of matter in a given volume).[1] By using this knowledge, one can upgrade middle-grade ore into a saleable concentrate by separation of heavier minerals (which tend to be valuable) from lighter ones, which are usually waste and can be discarded. This separation is often performed by running a suspension or slurry of the crushed ore through one of two types of "gravity" machine—a jig or a shaking table.[2] Precrushing plus "sizing" (by screening and/or other methods) is necessary before using either. Fragments to be concentrated in a large concentrator ("bull jig") are crushed into pieces no more than 1 inch in diameter. Finer crushing—down to 1/8 inch—is required if a shaking table is to be used. Pieces designated for the shaking table are screened to produce materials of classified sizes. The table is used to process particles that are approximately as large as match heads, but the "fines" and "slimes" are best removed and set aside for follow-up treatment with different types of equipment.

A bull jig can turn out a "clean" fluorspar concentrate as fast as a man or two can shovel coarse screened ore into it; a goodly percentage of the coarser fines can then be recovered by a smaller jig. The concentrate at that point is perhaps 95 percent pure fluorspar and 5 percent wall-rock or stray valueless particles. Shaking tables can do a reasonable job on the fines, but the lower "unit value" of fluorspar (as compared with precious metals), combined with the lower capacity

1. Some 99 percent of all minerals of value have much higher specific gravities (2.5 or so) than normal rock or earth.
2. A jig is a device that effects separation of valuable minerals from waste (gangue). The operation works by providing enough water to make a slurry of the crude ore. The separation is based on two complementary processes—"hindered settling" and upward flow of the slurry. Both processes facilitate a gravity-based separation and salvaging of the heavy particles.

of a shaking table all but eliminates using the latter machine for this application. However, even a small table can be economical for the recovery of gold from fine-ground ores.

A big shaking table, fed with classified feeds onto matching riffles,[3] will work best when the differences in specific gravity between the target mineral and the impurities are considerable. The best example I can cite was in the gravity mill of a high-grade gold mine near Pine Valley, California, that I was running before World War II (see pages 72-73). Its 2-ounce-gold-per-ton content put out a 1-inch-wide swath of very high value gold ore and a similar width of gold-bearing pyrite and other sulfides, which, although not nearly as rich as the high-value mineral, was still valuable enough to dewater and ship to the smelter. Between those two swaths there was a heavy sulfide mineral that had poor gold values; I judged it at the time to be an arsenide of some sort. In the course of preparation of this volume, I rethought this identification, and I think I now know what this impurity was (see Appendix II).

Because I had more mill capacity than mine production at Pine Valley, I assigned an older, experienced miner to produce extra mill ore by screening some of the "almost" ore that had been set aside in a small mine dump. The ore to be screened included 10-20% of waste rock easily identifiable "by eye" that could be discarded. This material failed to pass the screen and was discarded. The finer ore that had passed through the screen was then transported by truck to the mill's ore bin, which was less than 100 yards away.

It turned out that the mine-run ore carried excellent values in gold but also considerable amounts of arsenic. Smelters definitely do not hold a favorable view of arsenic. For one thing, the fumes from any of the chemical variants of this toxic element foul the atmosphere. Furthermore, arsenic interferes with gravity separation of the worthless molten slag from the much heavier molten gold, silver, lead, or copper ores, during their "treatment by fire." It tends to make the molten slag viscous, interfering with the free flow and settlement of the (heavy) valuable molten materials.

If we could have mined clean ore free from arsenic, we certainly would have done so, but we were committed to cleaning up our gravity concentrates by "selective flotation," after which we dehydrated our final concentrates for shipment to a lead smelter. If I remember correctly, we sent our ore by rail to the Asarco smelter at Selby, California, north of San Francisco.

---

3. Riffles are troughs having grooves or steps that catch gold particles.

APPENDIX II

I Believe the Arsenic Impurity at Pine Valley Was Lollingite

I never stopped wondering what the prominent arsenic mineral was that showed up on the table between the two valuable bands at the California gold mine I supervised in 1940. I was discouraged by my inability to find a specimen of the mineral larger than the head of a kitchen match. We had been taught in mining school to solve these problems one way or another, so I fired up an alcohol-fueled wick lamp and ran a flame test. To do this on a small specimen that is no larger than a "pinch," you have to use a blowpipe to concentrate the flame on the sample, which is laid into a shallow cavity on a block of charcoal. When I did this with the mineral from the table, a white sublimate appeared. This sublimate gave off a definite garlic-like odor; this odor confirmed the presence of arsenic.

At the time we were producing the two gold-containing bands (one of which had the arsenic impurity), I consulted my *Pocket Handbook of Minerals*. This classic text was written in 1908 by G. M. Butler, the Dean of the U of A School of Mines, and was updated in 1911.[1] I still use my copy of the second printing of that handbook. Set out in the "Sulfides" Section, I found lollingite and arsenopyrite (= mispickel); both are arsenides of iron. But I also found arsenides of nickel

1. Butler's book was a classic in its time, and it really still is. My copy is actually three books in one. The overall title is: *Handbook of Mineralogy, Blowpipe Analysis and Geometric Crystallography*. It contains *A Pocket Handbook of Minerals* (Butler, 1908; revised 1911); *Pocket Handbook of Blowpipe Analysis* (1910); and *A Manual of Geometrical Crystallography* (1918). It was the first of these that I actually used. It was written with, as the title page states, "little reference to chemical tests." It is a very early antecedent of modern field guides to minerals, except that it is really more thorough than recent books. I used Butler's book often, and many of the pages bear my handwritten notes. Minerals don't change much over the years, so for me, the book has never gone out of date. As useful as the book was to generations of geologists and mining engineers, I don't think it made a lot of money for Butler. The title page states: "Total Issue Four Thousand."

(smaltite) and cobalt (cobaltite). At the time I did my tests, I wasn't sure which mineral I had. But as I was preparing this manuscript I looked at the manual again. I had not taken into account the high specific gravity of our impurity. Lollingite has a specific gravity of 7.2, which is very close to that of our gold-table impurity, so now I believe that is what we had. I had made up my mind, I now believe incorrectly, that the impurity could be arsenopyrite. I knew that, even with the arsenide impurities, we had lots of good gold values in the concentrate that we shipped to the smelter.

That mineral—presumably lollingite—caused serious practical problems. The "firing order" of the five or six semiseparations produced by the flow of our table's riffled deck didn't separate the minerals well enough to make it possible to "cut out" relatively pure ore products, so we were forced to dewater and ship the whole "rainbow" to the smelter as "gold ore." I was told later that, after I had left to join the Army, the manager bought a small smelting furnace and smelted several tons of our table concentrates. However, the high percentage of the arsenic in the smelter's "charge" prevented clean separation of the molten slag from the molten "values" of lead and gold. It no doubt would have influenced the decision to attempt smelting had they known earlier what the impurity was.

APPENDIX III

Aiming Artillery Fire at Unseen Moving Enemy Ships

The version of the highly specialized art of hitting an unseen moving target that I mastered as "gun-pointer" at Fort Funston, California, when I was serving in Battery "B," 13th Coast Artillery, in the Harbor Defenses of San Francisco. dates from Spanish-American War times (i.e., the late 1890s). It depends heavily on the use of telephones[1] and super-accurate maps pin-pointing the relative positions of the seacoast cannons (or mortars); the observation posts; and, when the opportunity comes, enemy water-borne "targets of chance."

The harbor defenses of a seacoast port required setting up at least two observation posts sited on high, fortified places that commanded unobstructed views of the main channels of the seaport. Once this had been done and the relative positions of the two posts had been determined, the coordinates of each spot were determined and were plotted on a huge[2] map. The map existed for just one purpose—the guidance of our gunnery. It had fewer than two dozen key points on it. These points were vitally important to us but had no civilian value. Our map covered an area of 100 square miles. It was set up and maintained in a deep dugout. The map room needed no contact with "daylight" except for electric lines or batteries and telephones (which, of course, were on dedicated, restricted battery-operated systems). The phones were located at each observation post and at each gun commander's post. The gunner's and observer's posts, of course, had to be out in daylight, but the gunner might not even be within sight of the ocean.

---

1. It was the invention of the telephone that made this technology possible. In the 1870s, Elisha Gray and Alexander Graham Bell independently invented devices that became phones as we know them. They each rushed their designs to the patent office, but Alexander Graham Bell won the race by a matter of hours.. The two men entered into a famous legal battle over the invention of the telephone, which Bell won.
2. The map was at least the size of a queen bed.

The "hand" who actually manned our big map was a sergeant. During operations, he lay across the map in his stocking feet. He had a set of map scales he had prepared in India ink on scraps of muslin-backed paper to match the scale of the big map. He maintained connections with all the posts with portable military battery-operated phones.

When an enemy target was "sighted," it would still be out of range of the artillery "pieces." In practice, all the time we were there, we had no live targets to shoot at. But we had plenty of practice (sighting, not shooting!) on passing traffic, just for training. A lot of ships passed by our stations without realizing they were in our telescope's cross-hairs! We didn't fire on them, of course, but we figured how we could have done it. An electric signal-clock was "cut into" the phone circuit. This clock "dinged": "*-*—-*, *-*—-*" at 20-second intervals between the "single" dings, for the attention and guidance of all of the above-named personnel. The single ding following the gap gave us the order to fire.

The observers at each post had set the zeroes of the "dials" of their protractor-like theodolites to "Zero, on due North." When a target approached, they were given verbal telephone orders. For example, we might hear: "Black destroyer flying red-on-green flag, course left to right; reference point: base of stub mast amidships…commence tracking." The two observers immediately began tracking the given aiming-point on the moving target and called out their observed (horizontal) angle values, at 20-second intervals, to their map-attending counterparts plotting the data in the dugout.

Each of these dugout men operated a pivoted protractor-like nickel-steel "yardstick" divided into yards, like a ruler. The lines established by these instruments intersected each other so as to define the map points representing the target's aiming-point. The positions of these "rulers" were clamped while the plotter, atop the table, inserted a little two-piece metal replica of a tiny "flatiron"-shaped metal "Targ" inside the acute angle formed by the two "arms" and pressed down its little joined-on pointed end (all of this in one movement), while verbally ordering "Clear."

That "Clear" called for the immediate removal of both of the overlapping arms, and the plotter put a tiny hard-pencil triangle surrounding the tiny "nick" that the Targ's little punch had left in the (clean-paper) sheet atop the big map table.

By then, 20 seconds had passed, and the entire process was repeated at least half a dozen times, producing a virtually straight line that accurately depicted the course of the target. And, if all looked OK, the plotter selected the best-matching one of his set of "home-made" drawing-paper "scales" to use during the rest of this particular exercise. He may even have calculated the "knots" (minutes of latitude per hour) that each of his little pre-prepared scales depicted. Now, the point the Targ had just passed through was the first of the "set-forward" points to be plotted out ahead at 20-second intervals, for the artillery to aim at and fire at. (This meant that the projectile would land at a point where the ship would run into it, instead of vice versa.)

Either or both of the "yardstick" operators could read his scale of yards and give the master gunner and/or the plotter an idea as to the distance to the target. They needed to know the distance to give them an idea of how long it would take the projectile to reach the target. But they also needed to know the distance (and direction) from the "piece" to its assigned "aiming-point" on the target, predicted by the map's "set-forward-point" after each series of 20-second-apart single "dings."

In the meantime, the plotter had drawn an almost straight pencil line[3] that was an exact extension of the target's plotted course, and then he "scaled off" the distance the ship would go in the time the projectile was to be in the air. He penciled a circle around that—his "set-forward" point. Then another man—the "gun-arm" protractor operator—who manned a third yardstick-arm pivoted on the gun's map-position, took over momentarily. The plotter set the edge of the gun-arm alongside the set-forward point and read out the distance (in yards) that the gun crew needed, while the gun-arm operator read aloud the "azimuth" angle that his assigned protractor showed.

The gun-chief, who was almost always an officer, chalked up two angles (on a 4" x 16" blackboard that his gun-pointer could see): (1) the azimuth angle (given to him verbally by the gun-arm operator) and (2) the one he had to look up (in the gun's tailor-made operating manual)—the vertical angle required to propel the shell the proper distance to the set-forward point. This officer, who was sited at an overlook point not too far to the rear of the "piece" he was directing, had the authority to "correct" these two values at his discretion: He might have said "up 100 yards," "left so-and-so angle," or whatever. The latter command was probably given in "mils" rather than degrees/minutes.[4] First and foremost, of course, after attending to the proper setup, adjustment, and maintenance of the required materiel, was the training of the several-dozen-man teams, each member of which had to do his assigned split-second tasks, perfectly. *Every time*!

The whole thing sounds complicated, and it was! It used all of the experience and knowledge of geometry and trigonometry acquired beginning with Pythagoras. Some of the things we used were actually just standard surveying equipment. For example, my theodolite was my "buddy" as long as I worked for a living. I do have to say that after mastering all the procedures of our Coast

---

3. "A straight line?" you may ask. "Won't the target ship take evasive action?" "Yes, it will try, but a 20-knot ship can't change course fast enough to avert good gunnery."
4. A mil is one unit in 1,000; 20 yards at 1,000 yards is 20 mils.

Artillery that surveying was a snap. I didn't have to survey moving targets! Today, of course, coordinates are determined by GPS instruments. They make it easy to determine latitudes and longitudes. But hitting moving targets, however it is done now, must still require intricate equipment and judgment. It takes a lot of engineering to win wars and to defend our homeland, and this country has done pretty well with it.

APPENDIX IV

Building a Bridge the Right Way

In June 1944, before our unit (Company B, 1318th Engineers Regiment) was shipped overseas, I was placed in charge of building a 100-foot bridge at a stream crossing on a rural county road south of Charlotte, North Carolina. There were no plans, unless you call a 150-foot single-span of baling wire (which indicated where the center line was to be) a "plan." I was given about 75 enlisted men and turned loose. I had a supply chief, 2nd Lt. John Sankey, who was a very good right-hand man. If I had planned the bridge, I would have "reconned" upstream to see how high to make my piers to keep my timber bents[1] above flood level. However, I had all these men standing around, so I wasn't able to do much, if any, planning or designing. There is no way to work that many men in a small area such as a bridge site. The company commander must have had an oversupply of labor; he surely knew better! So I split my men into two teams. The morning team ate early breakfast and worked until late lunch. I ate breakfast with that group but then had an early lunch and worked the other half of my men until supper time, so each man worked only 5 or 6 hours, seven days a week. We had long June days and good weather, so it took only 16 days to finish the job.

I was lucky to find bedrock in less than knee-deep water, so in several days we were able to build forms and cast our concrete piers. These piers were built to support a single 40-foot span for which I had been given three new 42-foot 18-

1. A "bent" is a rigid frame (in my case made of 8" by 8" timber and set atop a concrete pier) oriented transversely with respect to the long dimension of the bridge and bolted down. The bents support the vertical loads of the beams and girders. My "bents" had to be very substantial because they would support both my stringers and my three steel I-beam girders, each of which I estimated weighed about one and a half tons and was approximately 42 feet long.

inch steel girders, plus nine 22-foot pine 6"-by-14" and 6"-by-16" stringers to span my three 20-foot remaining gaps.[2]

For equipment, we had a standard "government-issue" two-and-a-half-ton-truck-mounted air compressor with pneumatic pumps and jackhammers. We had a small Caterpillar tractor without a blade, and we had Lieutenant Sankey's assigned four-wheel-drive command car. Both the tractor and the car were mounted with power winches, each equipped with about 100-plus-foot cables. When we had the 8"-by-8" bents all in place, the company commander stopped by afoot, and my sergeant reported that he had seen him sight along the tops of the bents when I wasn't looking and burst into a broad smile. Of course, the tops had to be uneven because my stringers were of different heights. I don't think he knew that, and he thought he "had me" for sure. So I got ready to hoist all of the 22-foot-long timber stringers, which were to match my three steel girders. Each of those girders I estimated to weigh about one and a half tons.

The Battalion Commander "just dropped by" every day about 2 or 3 p.m. As I was mounting the bents (stringer supports), he asked, "Have you given thought as to how to hoist those steel girders into place?" "Not particularly, sir." "Do you know how to handle a gin pole?" "Yes sir." "Then use a gin pole."[3] And off he drove, as if he had "solved a major problem."

---

2. As I think about the bridge project today, a few things stand out in sharp contrast to the way things are done today. We used salvaged wood for the girders. In those days, it was common to salvage lumber. In fact, there were even lumber yards that specialized in used wood. In this case, the road was being realigned, and it was natural—then—to use material from the old bridge to build the new one. It was, after all, a big savings. Timber pieces of that size would have had to be shipped by railroad flatcar. Engineers today would cringe at doing things that way, but, as far as I know, the bridge didn't collapse. In fact, I built my home in Patagonia "around" six or seven salvaged 8"-by-16" former railroad trestle stringers. It hasn't fallen down, either. Another oddity—of course, it wasn't odd then—was that no cars came along. In those days, only the most well-to-do people owned cars, and in rural areas there weren't a lot of those.
3. A gin pole is an upright mast, which is guyed at the top to keep it in or near a vertical position. The pole would normally be timber; the main requirement is that it be strong enough to support the load being lifted. (For most jobs a telephone pole would do.) It can be hoisted by hand tackle or a power-driven hoist. It must be anchored by digging it in at least two feet, and it can be set up as high as 30 feet in the air. Well, to begin with, I didn't have a 40-foot pole. I had insufficient guywire, insufficient hold-fasts to secure the guywires, and less than a suitable footing for a gin pole. I had been familiar with gin poles as used to replace worn-out cannon barrels. Older cannon barrels always had cast-steel eyelets to which hoisting lines could be tied, plus a half dozen sets of block and tackles and experienced cannoneers to do the work, safely!

Really, in my judgment, he had not solved a problem but had created one. I still believe that a gin pole would not have been at all suitable for that job. I had no real holdfasts on my stringers to tie hoisting lines to. A gin pole, to have functioned at all, would have to have been a 30- to 35-foot timber pole and would have required a half-dozen block-and-tackle sets. I would also have needed a half mile or more of good heavy "line" (rope, in laymen's terms) and some noncoms with experience in using a gin pole of that size. In "Dutch" English, I've heard it said, "both of which I had neither."

Fortunately, a suitable way to hoist the girders had more or less invented itself. I was able to place the girders by putting a horizontal steel cable "highline" across the creek alongside the one-wire centerline I mentioned above. I sent a soldier up a living pine tree very near our assigned centerline with a lightweight line tied to his belt. When he got up about 20 feet, he hoisted up the hook end of our two-and-a-half-ton truck's hoist cable and wrapped it around the tree, which was about four and a half inches in diameter at that height. Unfortunately, the tree wasn't stiff enough at that height, so we sent another lad up the tree with another length of the same light line and had him wrap the hook of our dozer's hoist cable around the tree trunk, converting our two-ton dozer into a made-to-order "deadman" on the opposite (lower) bank, so that our "highline" would be guyed at both ends.[4]

Meanwhile, I had sent Lieutenant Sankey off to "scrounge up" an 8" or 9" cast-iron pulley. He was a good scrounger, and he found one. When he brought it back, we mounted it on our highline, fastened one end of the first girder to be placed to the pulley, and hoisted the stream end up with the winch on the front of Lieutenant Sankey's command car. We then lifted the "land" end up off the ground "by hand" (of a dozen men!) a little at a time.

You see, because the girders had been left there for us on the highest bank we needed only to move them laterally. So, with the pulley we lifted the streamward end of the girder 10 to 12 inches. In contrast, we lifted the landward end by hand only a few inches at a time. By moving the girder slowly but surely in this way, we were able to ease it into place on top of its assigned pair of bents. Once an object, no matter how heavy it is, is gravity free, it takes only a small "push" to move it laterally. In the absence of gravity, only frictional force has to be overcome. We had the first steel girder in place well before quitting time on the first day. There

---

4. A "deadman" normally is a holdfast with an emplaced bar holding a secured loop of a guywire; the bar is emplaced behind a notch so that tension on the guywire serves only to emplace the rod more firmly behind its immovable restraint.

was no assigned time for completion of the job, but I wanted to get the spans finished and the cables all reeled in before 2 p.m. of the second day—for sure.

The last thing I needed was for the Battalion Commander to find out that I wasn't using a gin pole, and it looked as if I was going to get by with it. With our system in working order, we knew we'd have no difficulty. On the a.m. shift, by 9 o'clock the next day, we had the second girder alongside the first, and by 10:30 a.m. we had the third one on the move. But here came the Battalion Commander, two or three hours early!! We were in the middle of setting the third and final girder, and because the men handling the steel would have been in jeopardy if we had stopped, we went right ahead and set the steel in place, spaced it properly, and started preparations to lay the flooring for the bridge. The major had very little to say.

In the private sector, solving a problem in a speedy, safe, and innovative way is usually rewarded. In the military, if engineers are ordered to "use a gin pole," they are supposed to use a gin pole. So basically I had disobeyed a direct order. The next day they sent a car for me to go back to camp and said my name had been drawn to act as "defense counsel" for an enlisted man's court-martial.[5] My men put the flooring on the bridge, and all manner of photos were taken to record the finished job. But all I got out of it was a hot shower. I never got a copy of the photos and, in fact, never even got back to take a look at the finished structure. However, I have the satisfaction to this day of having done the right thing.

Gin poles, if not properly operated, can be very dangerous, and had I tried to use one on this job, someone could have gotten killed.[6] As proper as I think my decision was, it followed me around for quite a while and certainly was reflected when I was awarded a mere "satisfactory" for that whole three-month period, which in military parlance is but one step away from a court-martial. However, placed in the same situation, I would have to do the very same thing again because it was the right way to do the job. Fortunately, I didn't have to question an order again during my hitch. So I got an "excellent" rating from then on, just as I had had before this incident.

---

5. My new "assigned job" was an easy one. I couldn't even find the nonexistent proceedings. There was no court-martial scheduled, so I moved back into my tent house in camp and took up my normal post duties.

6. Gin poles should be operated within a foot or less of vertical. They can't be safely swung "out of vertical" more than 5 degrees or so, but these steel girders were more than 40 feet long, and I had no holes or knobs to tie my "lift lines" to. This job was not a suitable place to use this often useful "tool."

Appendix V

The Last Train Ride

One of the most important features of old Patagonia was the railroad. Cattle and ore were shipped out regularly, and the mixed train[1] was an important means of going to Benson or Nogales. Today, we hop in our automobile, and in a short time we are in either of those towns. It was a bigger deal then. Much bigger. For one Patagonian, the late Ray Bergier (1916-2004), it was a very big deal indeed. He was born near the Sanborn Siding, now the Circle Z Ranch, between Nogales and Patagonia. They had to stop the train and rush his mother to the ranch house when she went into labor during a trip home from Nogales.

As I said earlier, a part of the line between Flux Canyon and Nogales was washed out in the summer of 1929. At that time the Flux-Patagonia section was repaired, and it continued to operate for two more years before being abandoned in 1931. In about 1935 the Calabasas—Flux Siding stretch of track was abandoned. Finally, in 1961, the last stretch, from Patagonia to Fairbank, was abandoned. The day of the railroad had gone. Today the old Depot sits between McKeown and Naugle Avenues, where it houses the town offices. The structure was moved 50 feet southward in 1964 to permit the new state highway to be widened through town.

In June 1961 the Pimeria Alta Historical Society took a last trip on the railroad from Fairbank (named for N. K. Fairbank, a railroad executive) to Patagonia and back. I prepared a log for the trip. That section was laid out in 1880 by a civil engineer, Raymond Morley, and was built shortly thereafter by the Santa Fe System. The dates may still be seen on the original steel rails (weighing 60 and 72 pounds to the yard) that are still visible in Patagonia.[2] This was the first railroad "main line" to enter Mexico from the United States when it was extended from

---

1. A mixed train had a baggage car plus a passenger coach. The car had two dozen or so seats. There was also a full-sized passenger car.
2. I have a dozen or more of the last rails; they are 30 feet long and were rolled in various foundries.

Benson to Nogales, via Calabasas, in 1882. The direct Tucson to Nogales connection was opened circa 1910.

Fairbank was the junction point for the now abandoned Bisbee Branch, also. On February 15, 1900, it was the scene of an attempted holdup of the eastbound train, by the Alvord gang. Burt Alvord had been a Willcox constable but had taken to crime. The robbery was foiled by Wells Fargo guard Jeff Milton, who killed one would-be bandit and seriously wounded another, although he was shot and left for dead himself. Milton recovered and later held mines in the Ajo region. I knew Jeff in later years; he had one badly damaged arm as a result of that robbery incident. Members of the gang were jailed, but Alvord escaped and was never recaptured. Some reports later claimed that he was in Canada or Central America (InformAction, 2004).

A partial log of the 1961 trip (with approximate mileages) follows. Points of interest not on the railroad itself are designated (S) south or (N) north:

Miles	Remarks
0.0	Fairbank, El. 3,860.
0.5	Bridge across San Pedro River.
1	Bridge across Babocomari ("caliche stalactite") Creek. Indian settlement (called "Santa Cruz" by Kino) is near here. Site of Quiburi is about 3 miles north.[3]
5	(S) Ruins of old stage depot.
7	Enter San Ignacio del Babocomari grant.[4]
11	(S) Huachuca Vista (Campstone Siding), El. 4,290.
14	(S) Camp Wallen ruins, 1866-69.[5]
18	High Bridge.
19	High ("Bee") Bridge. A swarm of bees has occupied this bridge for more than 40 years.

---

3. Along the San Pedro River near Fairbank are the ruins of a Spanish presidio (fort) known as Santa Cruz de Terrenate. Some archeologists think that the site had previously been occupied by a Sobaipuri Indian village called Quiburi. One archeologist, however, has suggested that the name probably refers to more than one location along the San Pedro and that the Indians kept the name as they moved to different locations (Negri, 2004).
4. The San Ignacio del Babocomari land grant, after much litigation, was approved by the Arizona courts. Of the 123,069 acres claimed, 33,792 was approved. The Babocomari Ranch essentially represents the approved acreage of the land grant.
5. This early Army camp eventually was merged into Fort Huachuca.

20	(S) Babocomari Ranch headquarters, on site of old Elias ranch house, circa 1835.
21	(N) Ruins of headquarters of old T-4 Ranch (clump of cottonwoods near track).
23.5	(N) Ruins of headquarters of old Pacheco Ranch.
24	Leave Babocomari Grant.
25	(N) Elgin, El. 4,720. Depot. In the movie *Oklahoma* the depot was labeled "CLAREMORE."
29	Leave valley of Babocomari Creek.
32	(N) Sonoita, El. 4,840. The railroad summit is near here. Site of former "wye" (to south) where helper engines were kept in steam-power days to help heavier trains up the steep grades to here from both east and west. Southwest of the box-car depot about 100 yards may be seen the grave of a railroad worker who was killed circa 1914 when he fell from the high timber bridge opposite Camp Crittenden.
32	Begin descent into valley of Sonoita Creek.
35	Small steel bridge.
35.5	(S) Old lime kiln operated in 1880 by John Smith to furnish "lime" for the Nogales lead smelter to use as flux. Smith is the father of Mrs. Charles May.[6]
36	High wooden trestle from which the worker fell. County road went under railroad here (eliminating a grade crossing) circa 1920.
36.5	(N) Ruins of Fort Buchanan (founded 1856, burned 7/21/1861 to prevent depot from falling into hands of Confederate Army). Ruins of Camp Crittenden, a post–Civil War camp (1867-72), lie about a half mile to the east.
37	(S) Cottonwood Spring, a warm spring with a good year-round flow.
38	(S) Rail X ranch headquarters, on site of old Pennsylvania Ranch, owned by Rollin R. Richardson (also—earlier—by Vail & Ashburn) in 1880s and 1890s.[7]

6. Charles May was an early resident of Crittenden. He lived alongside the tracks at the old lime kiln quarry.
7. From the mid-1940s through the 1950s, the Rail X Ranch was owned by the family of U.S. Representative Jim Kolbe.

41 (N) Casa Blanca, vanished site of town at mouth of Casa Blanca (White House) Canyon; had post office circa 1859 and a population of 50 in the 1870s census. First town in Sonoita Valley.

42 (N) Crittenden Station; stone house of Mrs. Charles May was built circa 1879. Site of lead smelter (see discarded slag west of May house, 1/4 mile).

44 Patagonia, El. 4,050. Founded circa 1898 as townsite of "Rollin" (by Rollin R. Richardson). P.O. established as Patagonia in 1900; incorporated February 16, 1948. The mission visita "Los Reyes de Sonoita" was set up southwest of here at various downstream locations in about 1697 by Father Kino.

For lunch and refreshment, stop at Patagonia; your engine (Diesel No. 5115), which has pulled the train known since 1881 as the "Burro," drops its number 944 and assumes the number 945 for the return trip to Fairbank and Benson. Check your tickets, watch your step, and be sure you are boarding the right train!

Additional Reading

Barnett, A. S. 1991. "The Pimeria Alta." In *Voices from the Pimeria Alta.* Nogales: Pimeria Alta Historical Society.

Canty, J. M., H. M. Coggin, and M. N. Greeley. 1999. *History of Mining in Arizona.* Vols. I-III. Tucson: Mining Foundation of the Southwest.

Dunning, C. H., and E. H. Peplow Jr. 1959. "Let's Talk About Mining," in *Rock to Riches.* Phoenix: Southwest Publishing Co.

Myrick, D. F. 1975. *Railroads of Arizona,* Vol. I. Howell-North Books.

Tuck, F. J. 1963. *History of Mining in Arizona.* Phoenix: Department of Mineral Resources. State of Arizona.

Voynick, S. M. 1978. *The Making of a Hardrock Miner.* Berkeley, West Virginia: Howell-North Books.

Williams, J. 1981. *From the Ground Up. Stories of Arizona's Mines and Early Mineral Discoveries.* Douglas: Phelps Dodge Corporation.

Young, O. E., Jr. 1974. *The Mining Men.* Kansas City, Missouri: The Lowell Press.

Young, O. E., Jr. 1976. *Black Powder and Hand Steel.* Norman: University of Oklahoma Press.

Young, O. E., Jr. 1976. *How They Dug the Gold. An Informal History of Frontier Prospecting, Placering, Lode-Mining, and Milling in Arizona and the Southwest.* Tucson: Arizona Pioneer's Historical Society.

Young, O. E., Jr. 1979. *Western Mining.* Norman: University of Oklahoma Press.

References

Many of the references used in this book came from the World Wide Web. We have included the URLs for such references, and, because we are aware that Web sites sometimes "disappear," we have also provided, in brackets, the search terms we used for the benefit of those who may wish to conduct their own Web searches.

Agricola, G. 1556. *De Re Metallica.* Basel: J. Froben and N. Episopius. Translated by Herbert Hoover and Lou H. Hoover. 1912. *The Mining Magazine.* London. Reprinted, 1950. Mineola, New York: Dover Publications.

American Forts Network. 2002. "Seacoast Forts of Hampton Roads." http://www.geocities.com/hrforts/Guns/guns.htm. ["Gran Puissance"]

Anonymous. 2003. "California: Palm Springs and Desert." American Journeys RV Travel Information. http://www.americanjourneys.com/. ["William P. Blake"]

Anonymous. 2004a. "Active 20-30 History." http://www.active20-30.com/history.htm. ["20-30 Club"]

Anonymous. 2004b. "All That Glitters." Fort Huachuca Museum historical file. Sierra Vista, Arizona: Fort Huachuca.

Anonymous. 2005. "155mm long GPF." http://users.belgacom.net/artillery/artillerie/5454.html. [Fillioux + gun]

Antiques Digest. 2005. "Watchman's Clock." http://www.oldandsold.com/articles02/clocks-wxyz.shtml. [watchman's clock]

Asbury, H. 1968. "The Great Illusion: An Informal History of Prohibition." Westport, Connecticut: Greenwood Publishing Group.

Barnes, W. C. 1983. Granger, Byrd (ed.) *Arizona's Names: X Marks the Place.* Scottsdale: Falconer Publishing Company.

Beck, A. M. 1998. "Burnside's Carbine & The Mowry-Cross Affair" http://www.civilwarguns.com/index.html. ["Sylvester Mowry"]

Bisbee Mining and Historical Museum. 2004. "The Bisbee Mining and Historical Museum." http://www.bisbeemuseum.org/. [Bisbee + mining]

Butler, G. M. 1911. *A Pocket Handbook of Minerals.* New York: John Wiley and Sons.

Butler, M. 2004. *Popular Piety and Political Identity in Mexico's Cristero Rebellion. Michoacan 1927-1929.* British Academy Postdoctoral Fellowship Monographs. Oxford: Oxford University Press.

California Parks. 2003. "California in Time: From the War With Mexico to Statehood." http://www.parks.ca.gov/pages/735/files/chronology%20from%201836%20to %201850%20statehood.pdf. ["Mormon Battalion"]

Calvert, J. B. 2004. "Old Units of Length." http://www.du.edu/~jcalvert/tech/oldleng.htm. ["English mile"]

Campbell, B. 2001. "Marines as Mail Guards." http://www.grunt.com/scuttlebutt/corps-stories/proud/mailguards.asp. ["train robberies" + Marines]

Campbell, T. E. 1939. *True Copy of the Original Notes of Thomas E. Campbell, Former Governor of Arizona in 1917.* Arizona Board of Regents Digital Library. http://digital.library.arizona.edu/bisbee/docs2/rec_camp.php. [Campbell + "Bisbee Deportation"]

City of Bisbee. 2004. "The Queen of the Copper Camps." http://www.cityofbisbee.com/bisb_history.html. ["Phelps Dodge" + copper + reserves]

De Camp, L. S. 1993. *The Ancient Engineers.* 1963. Reprinted by arrangement with Spectrum Literary Agency. New York: Barnes & Noble Books.

Dellachiara, M. 2004. "Lecture Is Worth Its Weight in Gold." The Daily Titan. http://dailytitan.fullerton.edu/issues/soring_02/04_26/news/lectureisworth.html. ["William P. Blake"]

DeMolay International. 2002. "What Is DeMolay?" http://www.demolay.org/whatis/. [DeMolay]

France Travel Guide. 2005. http://www.france-online.usa.sa.com/contact-us.htm. [Arles]

Geronimo, Turner, F., Barrett, S. M., and Turner, F. W. 1996. *Geronimo; His Own Story*. New York: Plume Books.

Gorby, R. 2000. "Alvar Nunez Cabeza de Vaca." Prescott: Sharlot Hall Museum Days Past. http://www.sharlot.org/archives/history/dayspast/text/2000_10_24.shtml. ["Baca Float"]

Hilton, G. W. 1994. *American Narrow Gauge Railroads*. Palo Alto, California: Stanford University Press.

Hinton, R. J. *The Handbook to Arizona*. 1878. San Francisco: Payet, Upham and Company. Reprinted, 1954. Tucson: Arizona Silhouette.

Holladay, A. 2002. "Shooting the Stars To Learn Where We Are." *USA Today*, March 27, 2002. http://www.usatoday.com/news/science/wonderquest/2002-03-27-sextant.htm. ["shooting the stars"]

InformAction. 2004. "Famous Cochise County Crimes." http://discoverseaz.com/History/Crimes.html. ["Alvord Gang"]

International Museum of the Horse. 1995. "The Buffalo Soldiers on the Western Frontier." http://www.imh.org/imh/buf/buf1/html. ["Buffalo Soldiers"]

Kasulaitis, M. N. 2005. "Arivaca Yesterdays: Sheep on the Arivaca Hills." *Connection*. April 2005. Arivaca, Arizona.

Keith, S. B. 1975. Arizona Bureau of Mines Bulletin 191, Index of Mining Properties in Santa Cruz County Arizona: 88 (Table 4); Schrader (1915): 2229-230; Rohrbacker (1964); Arizona Bureau of Mines file data. (Cited in American Boy Mine group, http://www.mindat.org/loc-34322.html.) ["American Boy Mine"]

Kitchel, D. 1948. "Fort Huachuca—Private Cattle Ranch or State Park?" Arizona Wildlife Sportsman. http://azwildlife.org/news/ summer2000/fthuac.htm. ["Camp Wallen"]

Lewis, S. 2004. "Bing Crosby's Lyrics."
http://www.kcmetro.cc.mo.us/pennvalley/biology/lewis/crosby/lyrics.html.
["Brother Can You Spare a Dime?"]

Luttwak, E. N., Ph.D. 2004. "Arms Control." Microsoft® Encarta® Online Encyclopedia. http://encarta.msn.com. ©1997-2004 Microsoft Corporation. All Rights Reserved. ["5-5-3 Treaty"]

McNamee, G. 1994. *Gila: The Life and Death of an American River.* Albuquerque: University of New Mexico Press.

Mabey & Johnson. 2005. "Mabey Logistic Support Bridge."
http://www.mabey.co.uk/johnson/LSB/index.htm. [Bailey Bridge]

Mendoza, J. (Joker). 2001. *Back in Them Days—When Patagonia Was a Mining Town.* Lincoln, Nebraska: Writers Club Press.

Miller, H. W. 1930. *The Paris Gun.* New York: J. Cape & H. Smith.

Mindat.org. The Mineral Database. 2005. "Patagonia District, Patagonia Mts., Santa Cruz Co., Arizona, USA." http://www.mindat.org/loc-4368.html. [Patagonia + Mining + District]

Mines Magazine. 2000. "Robert Peele's Mining Engineers' Handbook." Vol. 90, Number 2. Article reproduced at
http://www.alumnifriends.mines.edu/fun_stuff/peele/default.htm. ["Robert Peele"]

Montoya, M. E. 2002. *The Maxwell Land Grant and the Conflict Over Land in the American West, 1840-1900.* Berkeley: University of California Press.

NACCCA. 2004. "Roosevelt's Tree Army, A Brief History of the Civilian Conservation Corps." http://www.cccalumni.org/history1.html. [CCC + Roosevelt]

Negri, S. 2004. "Little Known Historic Sites: Where Early Man Killed Mammoths and Where the Apaches Drove Out the Spanish." Arizona Highways Online Extra. http://www.arizonahighways.com. [Quiburi]

Patrick, B. K. Adm. 2004. "David Glasgow Farragut."
http://www.military.com/Content/MoreContent?file=ML_farragut_bkp.
["Damn the torpedoes"]

Peele, R., and J. A. Church. 1941. *Mining Engineer's Handbook*. Vol. II. London: John Wiley & Sons, Inc.

Portillo, E. 2003. December 2, 2003. Tucson: *The Arizona Daily Star.*

Ricketts, N. 1996. *The Mormon Battalion: U.S. Army of the West, 1846-1847.* Logan: Utah State University Press.

Scorecard. 1997. Animal Waste Summary. http://www.scorecard.org/env-releases/aw/county.tcl?fips_county_code=04003#summary. [Cochise + sheep]

Serbian Orthodox Church..2004. "Listing of Serbian Orthodox Churches by Country." http://www.serbianorthodoxchurch.com/pages/listing/country/index.html. ["Serbian Orthodox Church"]

Spencer, Robert. 2000. "Play Ball!" Page 10: 1950s. The Charlottesville-Albemarle Historical Society. http://achs.category4.com/baseball/Play_Ball!_page_10__1950s.html. ["donkey baseball"]

Staff Writer. 2003. "Baseball Ain't About Horsing Around." Elysian Fields Quarterly—The Baseball Review.
http://www.efqreview.com/NewFiles/v20n/ortsider.html. ["donkey baseball"]

Stearn, J. 1989. *Edgar Cayce: The Sleeping Prophet*. New York: Bantam Press.

Taggart, A. F. 1927. *Handbook of Ore Dressing*. New York: John Wiley & Sons Inc.

Titley, S. R., and C. L. Hicks. 1996. "Copper Queen Mine, Queen Hill, Bisbee, Warren District, Mule Mts., Cochise Co., Arizona, USA." In *Geology of the porphyry copper deposits, southwestern North America*. Tucson: University of Arizona Press.

Todd, A. C. 1995. *The Cornish Miner in America*. Spokane, Washington: The Arthur H. Clark Co.

U.S. Federal Highway Administration. 2004. "Origins of the Interstates." Section 4. http://www.fhwa.dot.gov/infrastructure/origin04.htm. ["interstate highways"]

Virginia Dare Corporation. 1998. "History." http://www.virginiadare.com/. ["Virginia Dare" + Wine]

Weingroff, R. F. 2004. "Zero Milestone, Washington, D.C." http://www.fhwa.dot.gov/infrastructure/zero.htm. ["Bankhead Highway"]

White, M. 2003. "Source List and Detailed Death Tolls for the Man-made Megadeaths of the Twentieth Century." http://users.erols.com/mwhite28/warstats.htm. ["death tolls"]

White Pass and Yukon Railroad. 2004. "The White Pass and Yukon Railroad in World War II." http://drgw.free.fr/WP&YR/History/WWII/La2emeGM_en.htm. ["narrow gauge" + railroads]

Index

A

B

C

D

E

F

G

H

I

J

K

L

M

N

O

P

Q

R

S

U

V

W

X

Y

Z

978-0-595-67341-4
0-595-67341-4

Printed in the United States
37944LVS00004B/34-132

9 780595 673414